Marko Wolf

Security Engineering for Vehicular IT Systems

VIEWEG+TEUBNER RESEARCH

Marko Wolf

Security Engineering for Vehicular IT Systems

Improving the Trustworthiness and Dependability of Automotive IT Applications

VIEWEG+TEUBNER RESEARCH

Bibliographic information published by the Deutsche Nationalbibliothek
The Deutsche Nationalbibliothek lists this publication in the Deutsche Nationalbibliografie;
detailed bibliographic data are available in the Internet at http://dnb.d-nb.de.

Dissertation Ruhr-Universität Bochum, 2008

1st Edition 2009

Editorial Office: Christel A. Roß | Anita Wilke

Vieweg+Teubner is part of the specialist publishing group Springer Science+Business Media.
www.viewegteubner.de

Cover design: KünkelLopka Medienentwicklung, Heidelberg
Printing company: STRAUSS GMBH, Mörlenbach
Printed on acid-free paper
Printed in Germany

ISBN 978-3-8348-0795-3

Foreword

Information technology is the driving force behind almost all innovations in the automotive industry, with perhaps 90% of all innovations in cars based on digital electronics and software. Dozens of networked microprocessors and several hundred megabytes of software can be found in a common compact class car, controlling engine and driving functions, assisting the driver and enabling various comfort, infotainment and safety functions. One crucial aspect of digital systems in vehicles is their security. Whereas software safety is a relatively well-established (if not necessarily well-understood) field, the protection of automotive IT systems against malicious manipulations has only recently started to emerge. Even though many European car manufacturers have lately established R&D groups that are dedicated to embedded security in cars, so far there has not been an all-encompassing reference of this topic.

The book by Dr. Marko Wolf fills this gap, and is by far the most comprehensive treatment of IT security in vehicles available today. A particular challenge of automotive IT security is its interdisciplinary nature. Dr. Wolf has done an outstanding job incorporating disjoint areas in one comprehensive treatment. The book ranges from the relevant security technologies to a systematic analysis of security risks all the way to solution using state-of-the-art security methods.

Despite the fact that much of the material is based on results from the research community, Dr. Wolf succeeded in presenting all aspects in a very clear and accessible manner. I am convinced that the book will become an invaluable reference for designers and developers in the automotive industry, as well as for researchers in academia.

Prof. Dr.-Ing. Christof Paar
Chair for Embedded Security
Ruhr University Bochum

Preface

"I believe in horses. Automobiles are a passing phenomenon."
The German Kaiser Wilhelm II. (1859–1941) in 1905.

What the Motivation for This Book Is

In spite of the German Emperor's prophecy of doom, automobiles revolutionized our lives long ago. About a hundred years after the invention of the automobile, automotive technology itself is about to experience a revolution: the digital, networked car. Until twenty years ago, automotive vehicles were virtually closed electromechanical systems with only some small isolated non-critical IT applications, whereas today the fully software-driven, digitally networked and interactive vehicle is already in the offing.

Replacing steel and mechanics with information technology consisting of bits and bytes enables highly sophisticated, intelligent, interactive functionality, which could hardly have been realized solely with electromechanical controls. Furthermore, software-driven vehicular functionality improves flexibility as well as technical and economic efficiency considerably. By networking vehicles in such a way that they can wirelessly exchange messages with traffic infrastructures and other vehicles, passive dumb vehicles are changed into interactive intelligent communication nodes. Vehicular communications, for instance, will considerably improve road safety, if vehicles warn each other about local dangers. It would further help to face one of today's largest wastes of resources: the daily traffic jam. By enabling precise interactive real-time traffic control systems, traffic jams could be prevented before they actually set in.

This favorable development of vehicles becoming software-driven digital nodes within a vehicular communication network will, however, inherently introduce many dangers. Even though most vehicular applications are developed to face various (random) technical failures (e.g., by verifying checksums or ensuring high redundancy), they almost never consider a human attacker who uses a certain functionality in a syntactically correct way, but in a bad faith. Attacks on similarly complex and networked digital IT systems, such as personal computers, handheld devices, or web servers, can already wreak havoc. However, their malicious impacts are usually to some extent "limited" in terms of wasted time, lost sales, or even destroyed documents. The attack potential of malicious encroachments on vehicular IT systems, in contrast, goes from bogus warning messages over attacks on traffic IT infrastructures, which could cause the collapse of the entire traffic of a city, up to "terrorist attacks" on individual cars. For the latter, even simple encroachments on a real-time IT system, which controls two tons of steel at 130 km/h, can actually have devastating consequences for health and life of the vehicle occupants and other road users.

Despite these alerts, many of today's vehicular IT applications are susceptible to various malicious encroachments and require dependable security measures not only to ensure driving safety. The number of potentially endangered vehicular applications is astoundingly big. Even though mainly less critical topics such as antitheft protection or illegal chip tuning still dominate the area of vehicular IT security, many future applications, which have already left the desks of the automotive research departments towards series production, can never be realized without strong IT security measures. Dependable security measures are further essential to protect the liability, the revenues, and the expertise of vehicle manufacturers and suppliers besides various newly emerging security requirements (e.g., privacy) from drivers and occupants for several upcoming vehicular business and legacy applications.

The actual subject matter of this work is to prevent a similar scenario of ceaseless security vulnerabilities, as known from the world of personal computers connected to the Internet, in future vehicular IT systems. This is all the more important because in the automotive domain a single successful attack can already suffice to seriously jeopardize the public confidence in a brand, even if the actual endangerment remains marginal [Puc01]. The work you are holding in your hand is actually the first attempt to give a comprehensive and detailed insight into the emerging area of vehicular security engineering.

What This Book Is About

This work gives a comprehensive and detailed insight into the emerging area of vehicular security engineering, which aims to ensure the trustworthiness and dependability of vehicular IT applications. It should help you to understand the specifics whenever designing security critical vehicular applications while providing you with a solid set of general approaches, practical methods, and helpful implementation concepts. Therefore, it can be seen as a textbook as well as a practical guide, which helps you:

■ to learn about threats to vehicular IT systems by showing various current and forward looking exemplary security-critical vehicular applications,

■ to understand what causes these threats by analyzing possible attack incentives, attackers, and attack methods,

■ to reduce security vulnerabilities and security risks by providing practical methods for designing, implementing, and enforcing security efficiently in the automotive domain.

Thus, this book should, on the one hand, be of interest to automotive engineers and technical managers who want to learn about security technologies, and, on the other hand, to people with a security background who want to learn about security issues in modern automotive applications. In particular, this book can serve as an aid for people who need to make informed decisions about vehicle security solutions, and for people who are interested in research and development in this exciting field.

What This Book Is Not About

This book is not a replacement for reading one of those great security books that cover the topic of IT security in general. Even though this book tries to give a short general introduction into cryptography and IT security, it cannot replace further readings, which can be found for instance in [And01, SB07, Sti95].

This book is not concerned with vehicular IT safety. Even though IT safety and IT security are indeed interleaved fields and sometimes have fuzzy boundaries, this book is only concerned with the protection of vehicular IT systems against malicious encroachments (i.e., IT security) and not with precautions against random technical failures (i.e., IT safety).

This book is further not concerned with IT security of backend IT infrastructures. Protecting offboard servers and networks is a topic of its own, which is covered by general network and system security.

This book is also not a detailed implementation or configuration setup tutorial. This book intends to help you understand the specifics, which have to be taken into account whenever designing security-critical vehicular applications. Since IT security is rather a continuous individually tailored process than a standardized building block, virtually no book can give you a ready-made security solution which you could simply add to the corresponding product.

Finally, this book should not be useful to people trying to compromise vehicular IT applications or to break into vehicular IT systems.

How This Book Is Organized

This book is divided into three parts.

Part I: The Preliminaries. The first chapter gives a short introduction into the problem and helps you to understand and define the matter. It reviews related work in the field of vehicular IT security and provides a short introduction into security and cryptography, which is essential for vehicular security engineering.

Part II: The Threats. First, this part tries to raise awareness of the strong necessity of vehicular IT security by identifying, explaining, and classifying potential threats, potential attacks, and potential attackers for various current and future vehicular IT applications. However, this part also presents the multitude of new possibilities enabled by properly implemented vehicular IT security. This work further describes how to deduce appropriate vehicular security requirements which can thwart afore identified security threats properly. The last chapter of this part indicates characteristical advantages and constraints which arise when establishing IT security in the automotive domain.

Part III: The Protection. The final part of this work provides a solid set of practical security technologies and security mechanisms that are able to implement the identified security requirements efficiently and dependably in the automotive domain. Before ending with a detailed conclusion, it lastly describes some important organizational security aspects from the vehicular manufacturer's perspective, which have to be considered when establishing vehicular IT security.

Acknowledgments

Many people have helped me to realize this book. I wish to personally thank at least all the people from the Chair for Embedded Security at Ruhr University Bochum, in particular my Ph.D. advisor and highly valued mentor Christof Paar, my current employer escrypt GmbH who enabled me to continue working in my favorite research area in theory and practice, all the people from Vieweg+Teubner Verlag helping me to publish this book, and last but not least all my good friends, my loved family, and especially Ellen.

Marko Wolf

Table of Contents

II   The Threats   47

4   Security-Critical Vehicular Applications   49

List of Figures

List of Tables

List of Abbreviations

ABS	Anti-lock Braking System
ACC	Adaptive Cruise Control
AES	Advanced Encryption Standard
ASIC	Application-specific Integrated Circuit
C2C	Car-to-Car (communication)
C2D	Car-to-Device (communication)
C2I	Car-to-Infrastructure (communication)
CA	Certification Authority
CAN	Controller Area Network
CRTM	Core Root of Trust for Measurement
CRL	Certificate Revocation List
DES	Data Encryption Standard
DLP	Discrete Logarithm Problem
DoS	Denial of Service (attack)
DRM	Digital Rights Management
DSA	Digital Signature Algorithm
DSRC	Dedicated Short Range Communication
ECC	Elliptic Curve Cryptography
ECU	Electronic Control Unit
EDR	Event Data Recorder
ESP	Electronic Stability Program
EMSCB	European Multilaterally Secure Computing Base
FIPS	Federal Information Processing Standards
FPGA	Field Programmable Gate Arrays
GE	Gate Equivalent
GSM	Global System for Mobile Communications
GPS	Global Positioning System
GPRS	General Packet Radio Service
IEC	International Electrotechnical Commission
IP	Intellectual Property

IT	Information Technology
ITS	Intelligent Transportation System
IVC	Inter-vehicle Communication
IVU	In-vehicle Unit
LBS	Location-based Services
MAC	Message Authentication Code
MMI	Man-Machine Interface
NIST	National Institute of Standards
OBU	On-board Unit
OEM	Original Equipment Manufacturer
OS	Operating System
PC	Personal Computer
PIN	Personal Identification Number
PK	Public Key
PKI	Public Key Infrastructure
PUF	Physical Unclonable Function
RFID	Radio Frequency Identification
ROSI	Returns on Security Investment
RSA	Rivest Shamir Adleman
RSU	Road Side Unit
RTOS	Real-time Operating System
SAE	Society of Automotive Engineers
SK	Secret (private) Key
SO	Security Objective
SR	Security Requirement
TC	Trusted Computing
TCB	Trusted Computing Base
TCG	Trusted Computing Group
TPM	Trusted Platform Module
TSR	Traffic Sign Recognition
TSS	TCG Software Stack
TTP	Trusted Third Party
UMTS	Universal Mobile Telecommunications System
V2D	Vehicle-to-Device (communication)
V2I	Vehicle-to-Infrastructure (communication)
V2V	Vehicle-to-Vehicle (communication)

VANET	Vehicular Ad-hoc Network
VC	Vehicular Communication
VDA	Verband der Automobilindustrie (German association of the automotive industry)
VIN	Vehicle Identification Number
WLAN	Wireless Local Area Network

Part I

The Preliminaries

1 Introduction

This chapter provides the motivation for the necessity of vehicular IT security, outlines this work, and summarizes the corresponding research contributions.

1.1 Hope for the Best, Prepare for the Worst

Up to the late 1970s, vehicles were pure mechanical or at most electromechanical machines. As shown in Figure 1.1, with the adoption of the first electronic components during the eighties, today's vehicular components are mostly based on digital hardware and software, while in the future particularly software-driven vehicular components will gain central importance. Software-driven vehicular applications and services enable considerable cost reduction (e.g., due to code reuse, easy copying, large-scale application of standard hardware), weight reduction and hence less fuel consumption, and, in particular, enable sophisticated and novel functionality that is hardly feasible solely in hardware.

The costs for vehicular electronics are estimated to approach the 50% margin in vehicle manufacturing in 2015 [SW03], whereas up to 70% of the development costs are related to the software of the vehicular electronics [Bro06, Fri04]. Perhaps more importantly, information technology (IT) is the driving force behind most vehicular improvements so that today already more than 90% of all vehicle innovations are centered around software and electronics [Fri04]. The applications are realized as embedded IT systems and ranging from simple control units to full-fledged infotainment systems equipped with high-end processors whose computing power approaches that of current personal computers. In premium vehicles, one can find up to 80 processors that are connected by up to five different bus types and up to several 100 megabytes of embedded software providing more than 2000 individual functions [Bro06].

Not surprisingly, many classical IT and software development technologies are already well established within the automotive industry, for instance, hardware-software co-design, software engineering, software component re-use, and software safety measures. However, one important aspect of modern IT systems still has only little attention in the context of vehicular applications: *IT security*. IT security is concerned with the protection of IT systems against malicious manipu-

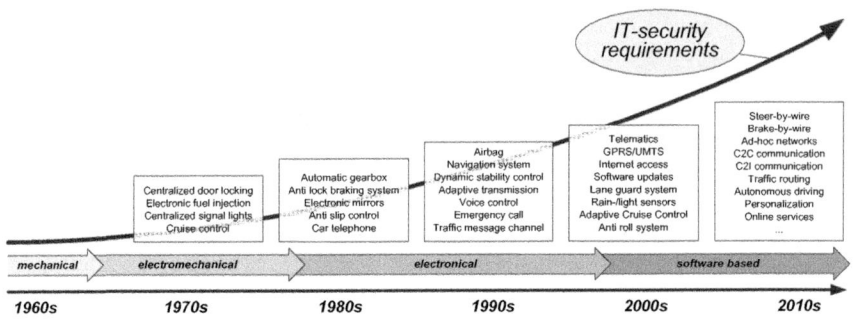

Figure 1.1: Development of vehicular applications from pure mechanical-based ones in the late 1970s up to almost complete software-driven vehicular applications in 2010 together with strong, continuously increasing need for vehicular IT security.

lations [ISO04a]. In contrast to *IT safety*, which provides protection against (random) technical failures (*fault-tolerance*), IT security according to [Shi00] means "measures taken to protect a system" in order to ensure necessary requisites that "system resources being free from unauthorized access and from unauthorized or accidental change, destruction, or loss." The differences between IT safety and IT security are once more depicted in Figure 1.2, which however also shows that both are required to build dependable vehicular IT systems. Nevertheless, IT safety and IT security are interleaved fields, that means, some technical failure (safety issue) can be used to realize some malicious threat (security issue) and vice versa. On the other hand, IT security measures can also help to improve IT safety and vice versa.

So far, there only existed niche applications in the automotive domain (e.g., immobilizers, digital tachographs) that particularly rely on security technologies. However, the situation has changed dramatically. More and more vehicular systems need security functionalities in order to reliably provide driving and road safety, and to protect the revenues, the liability, and the reputation of manufacturers and suppliers as well as to protect the various interests of the vehicle owner. Secure software updates for electronic control units (ECU), chip tuning prevention, mileage protection, and protection against counterfeit vehicular components are only some current examples for the need for vehicular IT security measures. Future vehicles will become even more dependent on IT security due to the following developments.

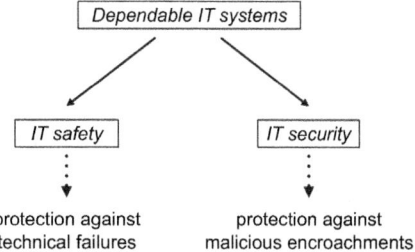

Figure 1.2: The differentiation between IT safety and IT security which, however, are both required to build dependable vehicular IT systems.

- Vehicles are increasingly controlled by electronics, where again software more and more determines most functionality and hence becomes the dominant factor [Fri04].

- Vehicular electronics are increasingly linked in (unprotected) internal networks that again will be linked with open unprotected external networks [PWW04c].

- Vehicular hardware and software becomes, manufacturer spanning, more and more standardized [Bro06] allowing possible encroachments to become more scalable and hence increasing potential incentives and the chances for a sufficient payoff.[1]

- Vehicle's anti-theft measures have to protect even individual components that are particularly valuable or safety-critical [HPWW05].

- Vehicles have to fulfill many new security-critical legislative requirements (e.g., secure emergency call [Eur08]).

- Vehicles will be involved in various new business models (e.g., aftermarket, pay-per-use, or location-based applications [RG06]).

- Vehicles will be exceedingly affected by counterfeiting [CNN07].

- Vehicles will communicate wirelessly with their surrounding infrastructures to enable new safety (e.g., electronic traffic signals), comfort and business applications [RPH06].

---

[1]Nonetheless, ongoing standardization certainly provides also some very valuable and indispensable characteristics that increase vehicular safety and security (cf. Section 6.3).

■ Vehicles will communicate wirelessly with each other; this will have an immediate impact on their driving operations, passengers's safety and privacy [RH07].

However, the majority of software and hardware in current vehicles is *not* protected against malicious encroachments. One reason is that previously, vehicles were merely closed electromechanical systems with only some small isolated non-critical IT applications, so that there were simply hardly any possibilities for worthwhile encroachments and hence only little incentives for OEMs to integrate proper IT security measures [MZ05]. Secondly, in many IT systems security often tends to be an afterthought, since achieving the core functionality first is a widely used approach even when designing a critical vehicular IT system. In fact, during the last 30 years most vehicular electronics were rapidly growing bottom-up, from simple, isolated, hardware-based dedicated microcontrollers to highly complex, interconnected, and software-driven distributed systems [Bro06]. Hence, a top-down design approach based on systematic software and security-engineering principles normally was never applied. Though, as can be seen for instance by the development of some operating systems or several Internet applications, implementing IT security afterwards is normally doomed to failure. Finally, it is complex and costly to establish the necessary organizational and technical measures and infrastructures for securing IT systems, which otherwise provides only little apparent promotional benefits for vehicle manufacturers and customers.

Nevertheless, IT security will play an important role for several future vehicular technologies and will even be an enabling technology for most future vehicular applications and IT-based business models. The target platforms within the vehicles that incorporate the security functions are *embedded systems*, rather than conventional general-purpose (desktop) computers. Some obvious differences in comparison to common general-purpose computer environments are listed in the following.

■ Embedded devices have small processors (often only 8 bit or 16 bit microcontrollers) which are limited with respect to their computational capabilities, memory, and power consumption. Hence, the application of common cryptographic primitives and protocols is limited just as well.

■ Embedded devices mostly have only limited possibilities and limited bandwidth for external communications. Hence, the extent and frequency of external communications, for instance, for key distributions, are comparatively small.

- Embedded devices are mostly very heterogeneous, widely distributed, and are assumed to be long-lived devices with only limited possibilities for any maintenance or updates.

- Embedded systems are often relatively cheap and cost-sensitive since they are often involved in high-volume products. Thus, adding a complex and costly security mechanism is seldom acceptable.

- Attackers of embedded systems have, in addition to the logical access, often also physical access to the target device itself that enables various additional attack techniques such as physical manipulations, side-channel attacks, or reverse engineering.

- Attackers of embedded systems have virtually unlimited time and unlimited trials to successfully mount an attack without having to fear to become detected.

- Attackers of embedded systems usually already have certain access authorizations or are legitimate users of the respective device they attack.

Thus, the technologies needed for securing vehicular applications mainly belong to the field of *embedded security* that fairly differs from the typical security problems and corresponding security solutions found for general-purpose computing and communication systems.

1.2 Outline

This work gives an insight into the emerging area of vehicular security engineering. Therefore, it is divided into three principal parts. After introducing necessary preliminaries (cf. Part I: *The Preliminaries*), it tries to raise awareness for the basic necessity of vehicular IT security by identifying and classifying potential threats for all kinds of current and future automotive IT applications. But, it also presents the multitude of new possibilities enabled by properly implemented vehicular IT security (cf. Part II: *The Threats*). It then describes how to deduce appropriate vehicular security requirements that can thwart identified threats properly followed by a detailed description of several security technologies and security mechanisms that are able to realize the identified security requirements within the automotive domain accordingly. This work lastly introduces several organizational security aspects and challenges that arise when establishing IT security in the automotive domain before it closes with a detailed conclusion (cf. Part III: *The Counteractive Measures*).

In doing so, this work first reviews *related work in the field of vehicular IT security* (cf. Chapter 2) and provides necessary *background knowledge in security and cryptography* required for vehicular security engineering (cf. Chapter 3). This introduction comprises a brief introduction into symmetric-key and asymmetric-key cryptography, introduces several cryptographic primitives and protocols and provides the basics in Trusted Computing and cryptanalysis. This chapter further gives some rough ideas for execution performance and required resources for cryptographic hardware and software implementations.

This work then identifies *security-critical vehicular applications* (cf. Chapter 4), including vehicular mechanisms and components, IT-based business models, legal applications as well as communication and safety applications that are current state-of-the-art, but also includes several future security-critical vehicular applications. Based on the identified critical vehicular application, potential *attackers and feasible attacks* (cf. Chapter 5) are identified and classified, followed by a detailed description of the steps involved in a *security analysis* in the automotive domain (cf. Chapter 6). It describes how to identify the particular security objectives of the entities involved in typical vehicular application and describes how to deduce the corresponding security requirements to fulfill the afore identified security objectives. It further indicates characteristical *technical and nontechnical constraints* as well as *characteristical advantages* that have to be faced while establishing IT security in the automotive domain.

Serving as base technologies to realize the identified security objectives and necessary security requirements, this work then provides an overview about general *vehicular security technologies* such as vehicular security architectures, physical security technologies, and vehicular security modules (cf. Chapter 7). For efficient and dependable implementation of the necessary security requirements, this work afterwards provides a solid set of *security mechanisms* (cf. Chapter 8) practically applicable in the automotive domain. IT security, however, always comprises both technical *and* organizational measures. Hence, this work furthermore gives an introduction into several important aspects of *organizational security* from the vehicular manufacturer's perspective (cf. Chapter 9), which have to considered when establishing vehicular IT security in the automotive domain. This work then closes with a detailed statement about *challenges and opportunities* for the automotive IT community for embedding IT security in vehicles (cf. Chapter 10).

1.3 Summary of Research Contributions

This work gives a comprehensive and detailed insight into the emerging area of vehicular security engineering that aims to ensure the trustworthiness and dependability of vehicular IT applications. Hence, it serves as a motivation, an introduction, and a deepening into most fields of vehicular IT security. Concretely, this work investigates the following research topics.

Security Threats in the Automotive Domain

The work starts with a state-of-the-art description of current developments in vehicular electronics while trying to raise an awareness and giving several reasons for the basic necessity of vehicular IT security. For that, it describes various vehicular mechanisms and components, IT-based business models, legacy and safety applications, which rely on dependable vehicular IT security measures. This includes security-critical vehicular applications that are very up-to-date such as vehicular software updates, but also several issues that are more forward-looking such as secure inter-vehicle communication. This work identifies, explains, and classifies potential threats, potential attacks, and potential attackers that are typical in the automotive domain.

This research contribution is based on the author's published work in [BEPW07, LPW06, PWW04b, WWW07].

Security Requirements Engineering in the Automotive Domain

This work also describes how to identify the individual security objectives of the entities involved in a typical vehicular IT application. It describes how to deduce the corresponding security requirements that fulfill the afore identified security objectives and can thwart all relevant security threats properly. For this, it moreover indicates some helpful advantages and several characteristic constraints that arise when establishing IT security in the automotive domain. This comprises also several organizational security aspects from the vehicle manufacturer's perspective.

This research contribution is based on the author's published work in [HSW06, PW08].

Vehicular IT Security Technologies and Security Mechanisms

This work further provides a solid set of practical vehicular security technologies and vehicular security mechanisms adapted for applications in the automotive domain that can implement the identified security requirements accordingly.

This comprises an overview about general vehicular security technologies such as physical security measures, vehicular security modules, and vehicular security architectures, but also concretely practical security mechanisms for vehicle component identification, secure vehicle initialization, vehicle user authentication, as well as cryptographic schemes for securing in-vehicle and external vehicle communications. For this, several solutions are based on the technology of Trusted Computing, which is well-established in today's PC world and newly emerges also into the world of embedded computing.

This research contribution is based on the author's published work in [BEPW07, HPWW05, HSW06, PWW04a, PWW04c, WWW07].

Intellectual Property, Expertise, and Software Protection in the Automotive Domain

This work lastly introduces feasible vehicular security architectures capable to enable several advanced schemes for intellectual property, expertise, and software protection. It describes new schemes for secure content distribution capable for—but not limited to—applications in the automotive world and in the world of mobile computing with its characteristic constraints. Therefore, it introduces new security protocols, components, and mechanisms based on the technologies of virtualization and Trusted Computing.

This research contribution is (partly) based on the author's published work in [AES+07, AOS+08, BEWW07, EGP+07, PWW05, SSSW06, SSW06, SSW08, WWW06].

2 Related Work

A great many of articles has been published, see for example [BFM$^+$03, CTG03, Dri02, EM00, HRS98, Moo90, Pol95, SV05, ZP93], that treat IT safety and reliability issues of vehicular electronics against random technical failures in detail. However, analyses that also consider malicious human manipulations, this means, that treat vehicular IT security issues, are still very rare. The few available publications are restricted to vehicular niche applications such as immobilizers [Bac97, Ulk00], the upcoming digital tachographs [And98], or vehicular closure mechanisms [SKC00] that actually employ security technologies in a limited and isolated manner.

However, together with the general advancement of digital vehicular electronics in the early 2000s (cf. Section 1.1), various authors have begun to identify the particular need for embedded security [KLM$^+$04, RKH04], vehicular software protection [Bro04, S$^+$02], and vehicular communication security [BE04, EZMTV02, HCL04, PP05]. Further publications regarding the necessity of vehicular IT security have pointed out vulnerabilities of existing security-critical applications [And03, FL06, LD04], have identified the need of system integrity protection [Ehl03, Rud03], and have identified several potential attackers in the automotive domain [Paa03].

In the following years, first practical proposals for establishing IT security in vehicles appeared. Such proposals provided, for instance, security requirement models and secure software update protocols [AHS05], mechanisms for run-time software attestation [SPvDK04] and component protection [GHOP01, HPWW05], secure electronic immobilizers [LSS05], internal [Fib04, PWW04c] and external [GGS04, RPH06] vehicular communication security, and firstly addressed privacy and content protection issues [DGL$^+$02, SHL$^+$05, Zim05]. Based on properly implemented vehicular IT security mechanisms, other authors identified new potential business models [PWW05, RG06] and possible safety applications [RCCL06].

The essential need for hardware-based security measures in embedded systems design has been identified early [AHSS05, vBC05, RKH04]. Thus, there already exist a few hardware-based approaches to include strong security also in embedded systems [AF04, Atm08, MMJ05, SLLG05, Tru05]. However, most are proprietary solutions and do not address the particular requirements and constraints within the automotive context (cf. Section 6.4.1). First publications for integrating special hardware for vehicular IT security not related to immobilizers or door locking mechanism appeared in [ER03]. This work first proposes the application of Trusted Computing (TC) technologies as developed by the Trusted Computing Group [Tru03] to solve current vehicular security issues. Further work [EHH⁺05, SSW06] then combined Trusted Computing mechanisms with virtualization technologies to enable embedded security architectures based on real-time capable minimalized operating system kernels (e.g., microkernels).

Today, vehicular IT security has become a lively field of study and is integral part of various vehicular IT related research consortiums [Car05, CVI04, eSa08, GST05, Net04, PRe04, SAF06, SeV06] and vehicular IT standardization projects [Aut03, HIS04, IEE06b, Jap08]. The first research workshop series that is explicitly related to vehicular security issues has been founded in 2003 [Emb03]. In the following years, several further related research workshop and conference activities followed [Aut04, Int04, Veh05, Veh04, Veh50], making vehicular security engineering a very recent and important area of research with several considerable real-life impacts [BDI⁺07, Kau06, Kuh06, MZ05].

An overview of recent publications in the main research areas of vehicular IT security: vehicular security hardware, secure car communication, vehicular software and IP protection, vehicular security architectures, component protection, vehicle access, and automotive privacy issues can be found in each corresponding chapter of this work.

3 Brief Background in Security and Cryptography

This chapter briefly provides necessary background knowledge in security and cryptography. It provides a brief introduction into symmetric-key and asymmetric-key cryptography, introduces several cryptographic primitives, and provides the basics in Trusted Computing and cryptanalysis. This chapter further gives brief descriptions of several cryptographic schemes that are important in the automotive context. It lastly gives some rough ideas for execution performance and required resources for cryptographic implementations in hardware and software. Parts of this chapter are based on published research in [AES$^+$07, EGP$^+$07, WWW07].

3.1 Enforcing the Secrecy of Secrets

Besides security enhancing technologies such as filtering (e.g., firewalls), anomaly detection (e.g., intrusion detection systems), or vulnerability scanning (e.g., antivirus software), the science of *cryptology* forms the base of information security. As shown in Figure 3.1, cryptology comprises two underlying information security technologies, namely, *cryptography* and *cryptanalysis*. Cryptanalysis is the science of codebreaking, which means methods for obtaining protected information without having access to the corresponding secrets, whereas *cryptography* refers to the science of information hiding, which means methods for protecting information against unauthorized access. Thus, cryptanalysis is inevitable to evaluate and assure the security of a cryptographic method. Cryptography, in turn, comprises cryptographic algorithms for information encryption and decryption based on a shared key (*symmetric-key algorithms*) or based on a public key and a private key respectively (*asymmetric-key algorithms*), hash functions, cryptographic protocols, and so on (e.g., use of cryptographic primitives as secure random number generators).

Understanding the basics of cryptology is essential for designing, analyzing, implementing, and assessing any IT-security system. Therefore, this chapter gives a short introduction into the basics of cryptography covering symmetric-key cryptography and asymmetric-key cryptography, cryptographic hash functions, mes-

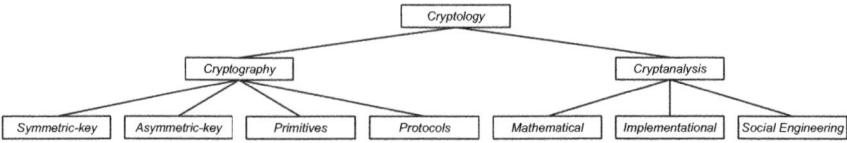

Figure 3.1: The various disciplines in the area of cryptology.

sage authentication codes, and cryptographic protocols relevant for vehicular security applications. Lastly, it gives a short introduction into the basics of cryptanalysis. However, further readings are helpful and can be found amongst others in [MvOV96, SB07, Sti95].

3.2 Symmetric-Key Cryptography

Symmetric-key cryptographic algorithms are the basic building blocks of any secure system that requires at least confidentiality. They are used to encrypt messages in bulk and to provide secure storage of data. These cryptographic algorithms use identical keys for encryption and decryption in both communicating entities and hence are called *symmetric ciphers*. Symmetric-key algorithms can be considered as a locked box with the messages inside that is sent to the other party. If one party has the right key to the lock, it can open and read all the messages in the box. The security of the symmetric cipher depends on the cryptographic strength of the key (the algorithm is assumed to be public). The exchange of these keys between the parties should be done using a secure channel, for instance, provided by an asymmetric-key cryptographic algorithm (cf. Section 3.3).

Symmetric-key algorithms can be mainly divided into two categories: *block ciphers* and *stream ciphers*. Block ciphers encrypt the messages in data blocks of fixed length, mostly 64 bits or 128 bits. The most popular block ciphers are the Data Encryption Standard (DES) [FIP77] and the Advanced Encryption Standard (AES) [FIP01]. The DES was the first standardized block cipher for commercial purposes and has been widely used because it was at the same time the only standardized and openly available algorithm that could be extensively studied by the cryptanalytic community. There have been no major weaknesses found (i.e., the DES is resistant against differential and linear cryptanalysis) in the algorithm to date to practically break it other than the relatively small size of the key that allows for a brute force attack running through all the keys. Today, a single DES key can be revealed already in less than 9 days, for instance, by performing an exhaustive key search with special-purpose hardware available for less than 10,000 Euros [KPP⁺06]. The DES finally expired as an US standard in 1999 and the National Institute of Standards (NIST) selected the Rijndael algorithm as the Advanced Encryption Standard (AES) in October 2000. In the transition phase, the Triple-DES was approved as a FIPS standard [FIP77]. The Rijndael algorithm [DR98] developed by Daemen and Rijmen was selected in an open challenge from a large set of algorithms submitted. The AES [FIP01] supports variable block and key sizes of 128 bits, 192 bits, and 256 bits to give a choice of different security levels based on its application. The AES has been optimized for efficient software and hardware implementations.

Unlike block ciphers, stream ciphers encrypt a plain text bit by bit. The most famous example is the one-time pad (OTP) [Ver26] encryption (also called Vernam cipher) which is the only known cipher which can be proven to be unbreakable [Sha49]. The OTP works by a bitwise XOR of the plain text with a one-time key, which is of the same length. The problem that a secret key of the same length as the message has to be transmitted over a secure channel makes OTP encryption inconvenient in practice. This shortcoming is overcome by using a pseudo-random generator as source for the secret key at the cost that the unconditional security does not hold anymore. Today's stream ciphers operate on a single bit of plain text (or a few bytes of data) being XOR-ed to a pseudo-random key stream that is generated based on a master key and an initialization vector. Stream ciphers are especially useful in situations where transmission errors are highly probable because they do not have error propagation. They also can be used when the data has to be processed one symbol at a time because of lack of device memory or limited buffering. Furthermore, stream ciphers often provide a higher throughput in comparison to block ciphers.

The most important symmetric-key (block cipher) algorithms that are currently employed in practice are the Data Encryption Standard (DES) [FIP77] and the Advanced Encryption Standard (AES) [FIP01]. Both are shortly introduced in the following.

3.2.1 Data Encryption Standard

As already mentioned in Section 3.2, the Data Encryption Standard (DES) [FIP77] is one of the most popular block ciphers and still used in many (embedded) applications and protocols, even though it is considered as insecure today. However, by successively encrypting a plain text p three times using three different encryption keys k_{1-3}, that means,

$$c = \text{DES}(k_1, \text{DES}(k_2, \text{DES}(k_3, p))),$$

the DES becomes (one variant of) Triple-DES or 3DES [FIP77], which has an effective security similar to key length of 112 bit [NIS07]. Thus, the 3DES is considered a very secure cipher and can be securely used also today.

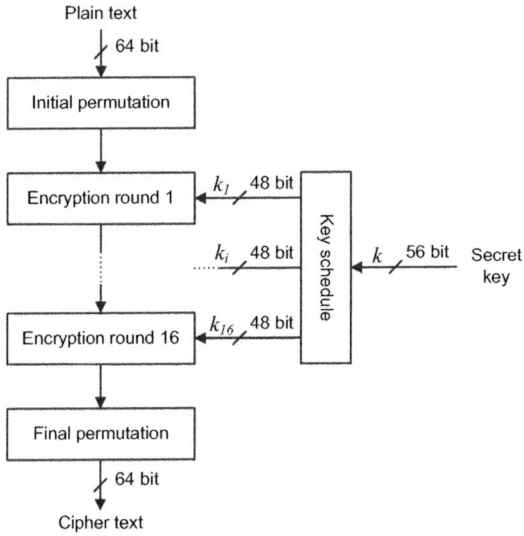

Figure 3.2: The iterative structure of one DES block encryption based on an initial permutation, 16 internal encryption rounds, and a final permutation. Every encryption round, in turn, uses an individual round key derived from the original DES encryption key.

As shown in Figure 3.2, the DES encrypts plain text blocks of 64 bits using an encryption key k of 56 bits[1]. After the initial bit-wise permutation of the 64 bit plain text, each block is successively encrypted 16 times. For this, each encryption round employs an individual round key k_i of 48 bits, which is derived from the original DES encryption key k by a key schedule algorithm. Each encryption round, in turn, represents a step within a so called *Feistel network*, which is depicted in Figure 3.3. Therefore, each block is split up into two equal halves of 32

---

[1] A DES key has an effective length of only 56 bits, since 8 bits of the full 64 bit DES key are just parity bits.

bit, where the right half R_i is fed into a special function $f(R_i, k_i)$ together with the round key k_i and the left half L_i becomes XOR-ed with the output of f. Lastly, the right and the left half are switched and put together again to a new 64 bit block $\{L_{i+1}, R_{i+1}\}$ for the next round with $L_{i+1} = R_i$ and $R_{i+1} = L_i \otimes f(R_i, k_i)$. After the 16th encryption round, both halves are switched once more and finally permutated again. The f-function, which takes R_i and k_i as input, internally again uses specific bit permutations, some non-linear functions (often called substitution boxes or S-boxes), and linear XORs to return a 32 bit output with a maximum of "confusion and diffusion" [Sha49]. Since the DES is based on a Feistel network, decryption uses virtually the same structure as encryption, except for the reversed order of round keys.

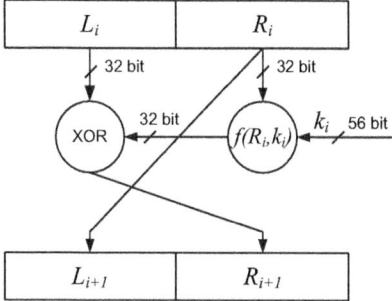

Figure 3.3: The Feistel structure of an internal DES encryption round.

Since the essential permutations, XORs, and non-linear substitutions of the DES can be implemented easily in hardware, there exist very fast, efficient, and quite small DES hardware implementations (cf. Section 3.7.2). Otherwise, permutations and non-linear substitutions can only be implemented moderately efficient in software. Thus, DES software implementations are comparatively slow and somewhat less efficient (cf. Section 3.7.1) than software implementations of other symmetric-key algorithms.

3.2.2 Advanced Encryption Standard

As indicated in Section 3.2, the Advanced Encryption Standard (AES) [FIP01] can be seen as the modern successor of the DES and hence is also widely used in various cryptographic applications and protocols. In contrast to the 64 bits of the DES, the AES encrypts plain text blocks of 128 bits using encryption keys of 128,

192, or 256 bits length[2]. Similar to the DES, the AES is very robust against known analytical attacks, in particular, against differential and linear cryptanalysis.

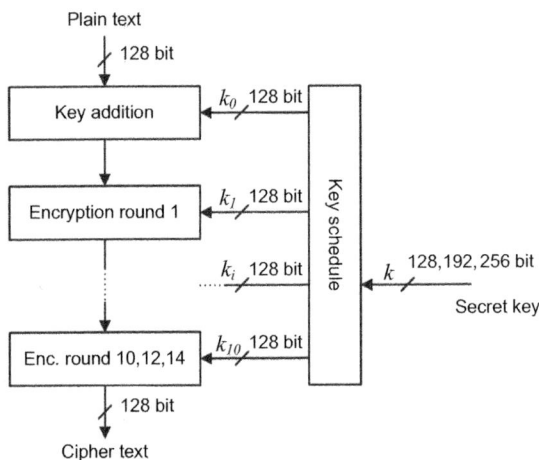

Figure 3.4: The iterative structure of one AES block encryption based on an initial key addition and—according to the actual key length—10, 12, or 14 internal encryption rounds. Every encryption round, in turn, uses an individual round key derived from the original AES encryption key.

As shown in Figure 3.4, the AES is not based on a Feistel network. The AES processes every 128 bit encryption blocks at once and *not* in two halves as the DES does. Instead, the AES is based on Galois field arithmetics. For each encryption round, the AES applies four different layers, namely the byte-substitution layer, the shift-rows layer, the mix-columns layer, and the add-round-key layer (cf. Figure 3.4). The byte-substitution layer represents the non-linear transformation (S-boxes) in order to increase the "confusion" and hence to increase the AES' resistance against known analytic attacks. The shift-rows and mix-columns layers, in turn, perform linear operations by permutating the block state on the byte level in order to ensure that flipping one input bit should change as many output bits as possible and hence to increase the "diffusion". Lastly, the add-round-key layer combines the current state with the round key, which has previously been derived

---

[2]The original Rijndael algorithm [DR98] provides block and key sizes of 128, 160, 192, 224 and 256 bits. However, the AES [FIP01] allows only a block size of 128 bits and 128, 192, or 256 bit encryption keys.

from the original AES key. As the round key has the same size as the block state (i.e., 128 bit), each byte of the round key is added with the corresponding byte of the block using bitwise XOR. Since the AES does not use a Feistel network, for decryption all layers have to be inverted and the order of round keys is reversed. However, the inverted layer operations for decryption are quite similar and only slightly slower compared to the layer operations used for encryption.

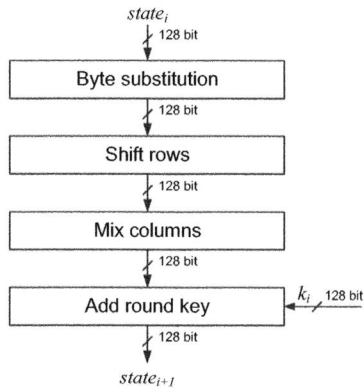

Figure 3.5: The four layers of an AES encryption round.

In contrast to the DES, the AES was developed to be efficient in hardware *and* in software. Nevertheless, in comparison to the DES, AES hardware realizations require some more hardware resources and achieve (somewhat) lower throughput values (cf. Section 3.7.2). On the other hand, optimized AES software implementations based for instance on so-called T-boxes [DR98] achieve very high throughput values that are at least three times higher than comparable DES software implementations (cf. Section 3.7.1).

3.3 Asymmetric-Key Cryptography

The main function of symmetric algorithms is the encryption of information, often at high speeds. However, there are two problems with symmetric-key schemes:

(1) Secure transmission of a secret key is required before being able to exchange messages.

(2) If in a network environment, each pair of users shares a different key, this will result in many keys[3]. Hence, this fact may result in problems handling the key management.

(3) After (secure) reception of a secret key, each party has to store its key securely for reuse.

The idea behind asymmetric-key or public-key cryptography can be visualized by making a slot into the locked box so that everyone can deposit a message (like a letter box). However, only the receiver can unlock the box and read the messages inside. This concept was first proposed by Diffie and Hellman [DH76] in 1976. Asymmetric-key cryptography is based on the idea of separating the key used to encrypt a message from the one used to decrypt it. Anyone who wants to send a message to another party, for example, to Bob, can encrypt that message using Bob's *public key*. However, only Bob can decrypt the message using his *private key*. It is understood that the private key should be kept secret at all times whereas the public key is publicly available to everyone. Furthermore, it is impossible for anyone to derive the private key from the public key (or at least to do so in a reasonable amount of time). With asymmetric-key algorithms one can realize at least three basic mechanisms, namely:

■ Key establishment and key exchange,

■ Digital signatures, and

■ Data encryption.

In general, one can divide practical asymmetric-key algorithms into three major families according to their underlying mathematical problem.

■ Algorithms based on the *integer factorization problem*: given a positive integer n, it is computationally hard to find its prime factorization. An algorithm based on the integer factorization problem is, for instance, Rivest-Shamir-Adleman (RSA) [RSA78].

■ Algorithms based on the *discrete logarithm problem* (DLP): given α and β it is computationally hard to find x such that $\beta = \alpha^x \bmod p$ where p is the respective modulus. Algorithms based on the DLP are, for instance, the ElGamal encryption system [ElG85] and its variant, the Digital Signature Algorithm (DSA) [FIP94].

---

[3]For a network with n users, $\frac{n \cdot (n-1)}{2}$ individual keys have to be shared before.

- Algorithms based on *elliptic curves* rest upon the DLP on the algebraic structure of elliptic curves over finite fields. Elliptic curve cryptography systems [Kob87, Mil85] are the most recent family of practical asymmetric-key algorithms, which have also gained wide acceptance including standardization [IEE00, IEE04].

There are many other asymmetric-key schemes, such as NTRU [HPH98] or systems based on hidden field equations, which are not in widespread use. The scientific community is only at the very beginning of understanding the security of such algorithms. Despite the differences between their underlying mathematical problems, all three algorithm families have something in common: they all perform complex operations on very large numbers, typically 1024 – 4096 bits in length for the integer factorization and discrete logarithm systems, and 160 – 256 bits in length for elliptic curve systems (cf. Section 3.4). This results in a poor throughput performance in comparison with symmetric ciphers. Nevertheless, asymmetric-key algorithms solve the key distribution problem in an elegant way, since the public part of the key can be distributed via an unsecured channel. Hence, one can establish a secure link between two parties without the need for an ulteriorly, previously exchanged secret. Thus, asymmetric-key encryption is normally used for transmitting only small amounts of data, like symmetric keys (cf. Section 3.9.2). Asymmetric-key algorithms are not only used for the exchange of secret keys, but also for authentication by using digital signatures. Digital signatures are analogous to handwritten signatures. They enable communication parties to prove to a third party that one party has actually generated the message, also called non-repudiation. The idea of the digital signature is appending a digital data block to the message that can be generated according to the message only by the person who signs it (like conventional signatures). Since the digital signature is a function of the message content and the private key, only the holder of the private key can sign the corresponding message. In practical terms, the private key is used for signing (thus only the holder of the non-public private key can sign a document) and the public key is used for the verification (thus everyone can verify the signature using the openly available public key). For practical implementations, using the RSA algorithm for digital signatures, a significant smaller public key[4] can be chosen to make the verification of an RSA signature a very fast and facile operation. Hence, RSA should be used in applications where the verification is done on an embedded platform and the signing on a personal computer or server. On the other hand, ECC should be used for applications where the embedded device performs signature generation as well as signature verification, since

---

[4]However, the private RSA key needs to have full length, for security reasons.

ECC is more efficient considering an application where the embedded device has to process both asymmetric-key encryption and asymmetric-key decryption.

The most important asymmetric-key algorithms are the Rivest Shamir Adleman (RSA) algorithm [RSA78], the ElGamal encryption system [ElG85], and elliptic curve cryptography (ECC) algorithms [IEE00]. The two most relevant asymmetric-key algorithms for the automotive domain, RSA and ECC, are hence shortly introduced in the following.

3.3.1 Rivest Shamir Adleman

In 1977, shortly after Diffie and Hellman's pioneering work in asymmetric-key cryptography [DH76], Rivest, Shamir, and Adleman proposed an asymmetric-key algorithm based on the mathematical problem of factorization of very large integers. Today, RSA is one of most popular asymmetric-key algorithms and is widely used in various cryptographic applications and protocols. Particularly, since the RSA patent has been expired in 2000, and it hence can be applied without any charges.

As with an asymmetric-key algorithm in general, RSA requires the computation of a public key PK_{RSA} and a private key SK_{RSA}. The public key PK_{RSA} can be revealed to everyone, whereas the private key SK_{RSA} has to be kept secret.

(1) Choose two distinct large random primes p and q.

(2) Compute modulus n as $n = p \times q$.

(3) Compute Euler's totient $\phi(n) = (p-1) \times (q-1)$.

(4) Choose a random integer e such that:

    a) $0 < e < \phi(n)$, and

    b) $\phi(n)$ and e are coprime, i.e., their greatest common divisor is 1.

(5) Compute the inverse $d \equiv e^{-1} \bmod \phi(n)$, that means, $d \times e^{-1} \equiv 1 \bmod \phi(n)$.

(6) Now, $PK_{RSA} = \{n, e\}$ and $SK_{RSA} = \{p, q, d\}$.

For RSA encryption, the message receiver (Bob) transmits its public key PK_{RSA} to the corresponding sender (Alice), but keeps its private key SK_{RSA} secret. The sender (Alice) then converts the plain text message in a set of numbers $M = \{m_0, \ldots, m_j\}$ with $m_i < n$ as defined in corresponding *padding schemes*. A feasible padding scheme for RSA is the Optimal Asymmetric Encryption Padding (OAEP) as specified in the Public Key Cryptography Standard PKCS#1 [JK02].

For message encryption, the sender (Alice) then computes for all m_i the corresponding $c_i = m_i^e$ mod n and hence transmits a cipher text $C = \{c_0, \ldots, c_j\}$ to the receiver (Bob). For RSA decryption, the message receiver (Bob) uses his private key SK_{RSA} to decrypt C back to M via $m_i = c_i^d$ mod n. Lastly, the receiver (Bob) converts M to the plain text message again, using the RSA padding scheme both parties agreed upon.

RSA can be implemented quite efficiently in software (cf. Section 3.7.1) and in hardware (cf. Section 3.7.2). Particularly, RSA encryption operations and hence likewise RSA signature verifications can be implemented very efficiently by choosing a small public exponent e. This means, if the private exponent d has the full length of n and a proper padding scheme is used as well (i.e., one that embeds randomness), the public exponent e can be a very small integer, such as $e = 3$ [Hås85], which, in turn, increases the encryption performance considerably. Nevertheless, RSA operations are by orders of magnitude slower than any symmetric-key operation. Thus, RSA operations are mainly used for exchanging secret keys (cf. Section 3.9.2) or digital signatures (cf. Section 3.9.1).

3.3.2 Elliptic Curve Cryptography

As already briefly mentioned in the introduction of asymmetric-key algorithms, algorithms based on elliptic curves are the most recent family of practically applied asymmetric-key algorithms that were firstly[5] and independently proposed by Miller [Mil85] and Koblitz [Kob87] in the late 1980s. In the late 1990s, elliptic curve algorithms were proven to provide a similar (or even slightly higher) level of security as the two other practically important asymmetric-key families, which are based on the integer factorization (e.g., RSA) and the discrete logarithm (e.g., ElGamal). Moreover, in contrast to RSA and ElGamal, ECC systems use considerably smaller keys, provide some performance advantages, and have some lower demands on memory and bandwidth. Lastly, only RSA encryptions (and hence also RSA signature verifications) based on small exponents (cf. previous section) are faster than usual ECC encryptions. Nevertheless, ECC employs a slightly more complex arithmetic and is subject to various patents (but not completely).

Figure 3.6 shows an exemplary elliptic curve over real numbers, whereas elliptic curves used in cryptography are defined over so-called *Galois fields*, which roughly means a finite set of elements with a well-defined set of corresponding arithmetic operations. Typically, ECC systems are defined either over a prime Galois field $GF(p)$ with p being a large prime, or over a binary Galois field $GF(2^m)$,

---

[5]However, the algebra and geometry of elliptic curves are already known for over 150 years.

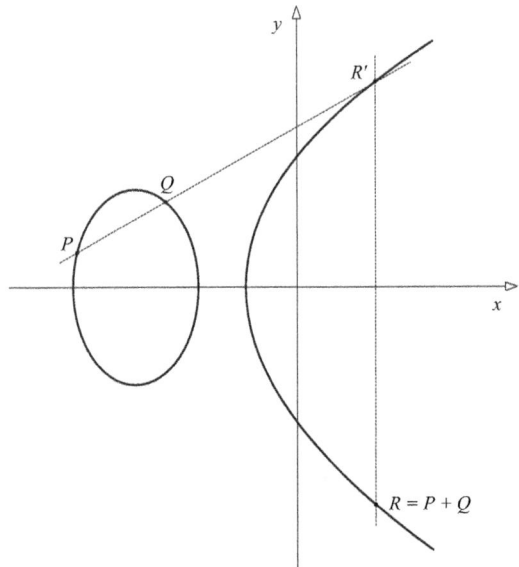

Figure 3.6: Geometrically, $P+Q$ is the inverse of the point R' that lies on the intersection of the line through P and Q and the elliptic curve.

which both apply modular arithmetics accordingly. Concretely, an elliptic curve over $GF(p)$ refers to the set of all points $P = (x,y)$ that fulfill the equation:

$$y^2 \equiv x^3 + ax + b \bmod p$$

where a and b are constant values in $GF(p)$ so that $4a^3 + 27b^2 \neq 0$. An elliptic curve over a binary Galois field $GF(2^m)$, in turn, is the set of all points $P = (x,y)$ that fulfill the equation:

$$y^2 + xy \equiv x^3 + ax^2 + b \bmod 2^m$$

where a and b are constants in $GF(2^m)$ and $b \neq 0$. Further, elliptic curve cryptography systems define specific operations between two points P and Q on a curve, such as the addition of two points (e.g., $P + Q = R$) or the multiplication of a point with a constant integer k (e.g., $Q = kP$). Figure 3.6 shows the addition of two points on an elliptic curve geometrically, such that adding the two points P and Q on the curve results in a third point R also on the curve. However, already small variations on the positions of P and Q can cause huge variations in the resulting position of R. The particular algebraic definitions and many further readings regarding ECC can be found amongst others in [HMV04, Was04]. However, the underlying mathematical difficulty of an elliptic curve cryptography system can be stated as follows. Given two points P and Q on an elliptic curve ec, it is "hard" to find an integer k such that $Q = kP$ (if k exists at all). This is also called the elliptic curve discrete logarithm problem (ECDLP). Therefore, given for instance an elliptic curve ec over $GF(p)$ with the order of n, an ECC-based asymmetric-key cryptographic system can be built as follows.

(1) Choose a point P on ec.

(2) Choose a random integer k with $0 < k < n - 1$.

(3) Compute $Q = kP$ by applying multiple additions of P.

(4) Now, $PK_{ECC} = \{ec, P, Q\}$ and $SK_{ECC} = \{ec, k\}$.

Since in fact, elliptic curve cryptography is based on the generalization of the discrete logarithm, ECC can be adapted to realize also various DL-based cryptographic schemes. Thus, ECC can be used to realize various encryption schemes, key agreement schemes, and digital signature schemes such as the:

- Elliptic curve integrated encryption scheme (ECIES),

- Elliptic curve Digital Signature Algorithm (ECDSA),

- Elliptic curve Diffie-Hellman key agreement scheme (ECDH), and

- Elliptic curve Menezes-Qu-Vanstone key agreement scheme (ECMQV),

which are all specified in [ANS95b, IEE00, IEE04]. As mentioned before, ECC is believed to replace DL-based cryptography and RSA cryptography at least in the embedded world. However, the underlying elliptic curves have to be chosen carefully, since at least certain elliptic curve families are considered to be cryptographically weak (e.g., supersingular elliptic curves).

Even though ECC hardware and software implementations are currently somewhat rare, there already exist various ECC implementations for desktop computers and for the embedded world (cf. Section 3.7). In particular, ECC implementations are faster than generic RSA implementations (i.e., RSA implementations not using small exponents for encryptions) of the same security level. Nevertheless, ECC signature verifications are by a factor of $2 - 5$ slower than ECC signature generations. Moreover, in comparison to RSA signature verifications using small exponents (cf. the previous section), ECC signature verifications are in orders of magnitude slower. Thus, ECC should be used in cryptographic applications that require both signature generation and signature verification.

Lastly, there exist patents on ECC on various levels, which, however, mainly cover specific implementation methods, specific Galois field structures, and several countermeasures against side channel attacks. Hence, ECC can be realized efficiently and securely without affecting any patents, but nonetheless often involve a careful patent search.

3.4 Recommended Key Lengths

The recommended key lengths for cryptographic algorithms particularly depend on the different power of the corresponding attacks and the projected protection time-frame. Since for most symmetric-key algorithms currently brute-force attacks are the best known attack, for asymmetric-key algorithms exist considerably more powerful attacks than successively trying all possible keys (i.e., analytical attacks). Thus, the recommended key sizes for asymmetric-key algorithms are normally considerably larger than symmetric-key lengths.

	AES/DES	ECC	RSA
Short-term security	64 bit	128 bit	700 bit
Middle-term security	80 bit	160 bit	1024 bit
Long-term security	128 bit	256 bit	4096 bit

Table 3.1: Recommended key lengths for short-term, middle-term, and long-term security for some selected symmetric-key and asymmetric-key cryptography algorithms.

Table 3.1 puts the asymmetric-key and symmetric-key bit lengths in perspective. This recommendation assumes that in the near future there will be no unexpected mathematical attacks. The three temporal classifications refer to the time-frame for that a secret information can be assumed to be protected. Hence, short-term security refers to a few days, middle-term security to a decade and long-term security to several decades, all without the eventuality of quantum computers. Even though quantum computers are still in their infancy, they have potential impacts in cryptanalysis. While most symmetric-key schemes can be protected against quantum computer cryptanalysis by choosing key lengths of 256 bit and above, asymmetric-key algorithms based for instance on integer factorization (e.g., RSA) or the discrete logarithm (e.g., ElGamal) would quickly become insecure [Sho97]. However, choosing the appropriate key length depends much on the kind and security targets of the respective application. Highly security-critical vehicular applications such as digital tachographs, motor control units, or immobilizers have to provide *at least* middle-term security, whereas less security-critical applications such as personalized presets or customer information services could apply even short-term security. Although OEMs hardly provide any public information about applied security standards, at least two useful references providing key length recommendations are [MKK+06] for vehicular flash security and [IEE06b] for wireless vehicular access.

3.5 Hash Functions

Hash functions "compress" a digital information of any length to a (nearly) unique string of fixed length, the so-called *hash value* or *digital fingerprint*, which is the unique representation of the corresponding digital information. In cryptography, hash functions are used in many applications, for example, digital signatures, pseudo-random number generators, message authentication codes (e.g., HMAC).

As shown in Figure 3.7, a hash function divides the input information x into blocks of equal length x_i (i.e., typically 512 bit) that are sequentially processed by a non-reversible compression function, which finally returns the hash value $y = h(x)$. This iterative processing structure is also known as *Merkle-Damgard construction*. Hash functions are *one-way functions*, that is, for (almost) all given outputs y, it is impossible to find any input x such that $h(x) = y$. Hence, with a given input, a hash value can be computed, but it is computationally infeasible to compute the input if only the hash value is known. A collision-free hash function is a hash function where an attacker cannot find two inputs x_a and x_b that compute the same hash value y. Since hash functions map more than one value to the same hash, a collision cannot be prevented. Thus, hash functions have to provide at least *strong collision resistance*, that means, it is computational infeasible to find two different inputs x_a and x_b that compute the same hash value $y = h(x_a) = h(x_b)$. For *weak collision resistance*, it is computational infeasible to find two inputs x_a and x_b that compute the same hash value y in case x_a and corresponding $y = h(x_a)$ are already given.

Even though the basic principle of an error-detection code is quite similar to that of a cryptographic hash function, error-detection codes cannot be used for computing cryptographic fingerprints. Error-detection codes are designed to detect random errors (i.e., to provide safety), however, they cannot prevent malicious ma-

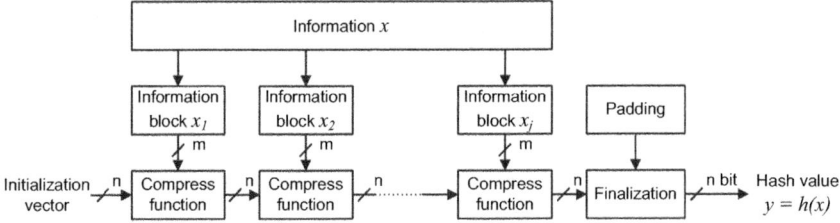

Figure 3.7: Merkle-Damgard construction of a hash function that divides the input information x into j blocks of of equal length x_i (that includes proper padding) to become sequentially processed by a compression function.

nipulations (i.e., provide security). The difference between error-detection codes and cryptographic hash functions can be clearly exemplified, using the digital signature scheme (cf. Section 3.9.1) that normally signs the digital fingerprint of an information, as an example. If here an error-correction code is used for computing the signed digital fingerprint, one could easily create an identical fingerprint for a completely different information, which would not be possible with a cryptographic hash function.

Nowadays there are several families of hash functions. The MD-family [Riv92] and SHA-family [FIP02b] are the ones mostly used. The MD-family generates hash values up to 128 bit, but suffers from serious flaws[6] making further use of the algorithm for security purposes questionable. The SHA-family was developed by the NSA in 1995 (updated last in 2004) and generates hash values up to 512 bit. Attacks have been conducted also within the SHA-family, particularly for the widely used SHA-1 (160 bit hash value). No attacks have yet been reported on the higher SHA variants (256 bit and 512 bit), but since they are similar to SHA-1, researchers are worried, and are currently developing candidates for a new hashing standard [Nat08]. Nonetheless, hash functions can be also constructed by using symmetric block ciphers with a constant non-secret key [MvOV96].

Hash functions can be efficiently implemented in hardware and software (cf. Section 3.7). However, at least for SHA-1, the hardware throughput improvements compared to the software throughputs are rather small in comparison, for instance, with DES or AES hardware throughput improvements. Nevertheless, SHA-1 can be implemented on typical desktop computers with throughput values of several hundred Mbit/s and reaches Gbit/s on dedicated hardware implementations.

---

[6]There exist algorithms to find a collision within minutes using a standard computer [LL07].

3.6 Message Authentication Codes

In contrast to keyless hash functions (cf. Section 3.5), message authentication codes (MACs) inherently involve a secret symmetric key. Thus, MACs can provide information integrity and information authentication in all security scenarios, where the involved entities are able to share a secret key. Then, MACs are able to provide several (certainly not all) security properties that are also provided for instance by digital signatures (cf. Section 3.9.1), while being considerably faster than digital signatures. Hence, MACs are sometimes also referred as *symmetric signatures* or *cryptographic checksums*. The basic steps involved in generating and verifying an authentication tag t_M for an information M are as follows.

(1) The *sender* and the *receiver* of an information M agree upon a random secret symmetric key k and securely exchange k while each entity keeps k secret.

(2) The *sender* uses a message authentication code MAC and the shared secret key k to produce an authentication tag t_M for M such that $t_M = \text{MAC}(M, k)$.

(3) The *sender* sends both t_M and M to the *receiver*.

(4) The *receiver* uses the same message authentication algorithm MAC and the previously shared secret key k together with the received M and individually computes an authentication tag $t'_M = \text{MAC}(M, k)$.

(5) The *receiver* accepts the corresponding information M only if the individually computed authentication tag t'_M and the received authentication tag t_M are identical.

Thus, a message authentication code provides the following security properties:

- Only holders of the secret symmetric key k are able to generate a proper message authentication tag.

- Only holders of the secret symmetric key k are able to verify a message authentication tag accordingly.

- The information and the message authentication tag are directly related, that means, a (valid) message authentication tag cannot be reused to authenticate another information.

However, in contrast to digital signatures, a message authentication cannot provide non-repudiation. Since each entity that has access to the shared secret k is able to create valid message authentication tags, message authentication codes cannot protect two entities against each other. MACs can be built from block ciphers and hash functions. Hence, in the following, a brief introduction into both possible realizations is given.

3.6.1 Block Cipher Based Message Authentication Codes

Message authentication codes based on symmetric block cipher algorithms (cf. Section 3.2) normally employ the respective algorithm using the cipher block chaining (CBC) mode of operation. Thus, for generating a message authentication tag t_M, the information M is divided into j blocks of equal length compatible with the underlying cipher algorithm (e.g., 64 or 128 bit) such that $M = \{m_0, \ldots, m_j\}$. After the first information block m_0 has been encrypted using the shared secret key k, all remaining blocks m_i are then XOR-ed with the preceding encryption result y_{n-1} and subsequently encrypted using k as follows.

$$t_M = y_j = \forall i [\mathrm{Encrypt}(m_i \otimes y_{i-1}, k)]_{i=1\ldots j} \quad \text{with} \quad y_0 = \mathrm{Encrypt}(m_0, k)$$

Thus, the last encryption result y_n represents the message authentication tag t_M. For the verification of the message authentication tag t_M, the receiver applies the same procedure and compares its individually computed authentication tag t'_m with the received authentication tag t_m. In case $t_m = t'_m$, the information is properly authenticated and verified for integrity. A block cipher based message authentication code using the DES algorithm (cf. Section 3.2) has been standardized in [ANS95a].

3.6.2 Hash Function Based Message Authentication Codes

Message authentication based on hash functions can take advantage of the particular computational efficiency of hash functions (cf. Section 3.5). Thus, hash-based MACs are considerably faster than MACs based on block ciphers. The most popular and widely used hash based MAC is the keyed-hash message authentication code (HMAC), which has been standardized in [FIP02c] and can be used with any hash function. Hence, the efficiency and the cryptographic strength is defined by the underlying hash function. Moreover, if the underlying hash function fulfills the assumptions defined in [FIP02c], the HMAC can be proven to be secure. However, for generating a message authentication tag t_M based on a hash function Hash, the corresponding HMAC equation results in the following.

$$t_M = \mathrm{Hash}\{(K \otimes opad) \| \mathrm{Hash}[(K \otimes ipad) \| M]\}$$

The secret key k is first padded with zeros to obtain K such that K has the length of a hash block (e.g., 512 bit). Then K is bitwise XOR-ed with the two one-block-length hexadecimal constants for outer and inner padding $opad = 0x5c5c5c\ldots5c5c$ and $ipad = 0x363636\ldots3636$ respectively. Now M is appended to the result of $K \otimes ipad$ and the resulting concatenation is hashed using Hash. Lastly, the resulting hash value is appended to the result of $K \otimes opad$ and hashed once more using Hash.

3.7 Cryptographic Implementations

In the following, some exemplary software and hardware implementations of the cryptographic primitives introduced in this chapter are presented. However, since most implementations particularly depend on the respective application and the actual computational environment, this section can give only rough ideas about expected performance values and required hardware and software resources.

3.7.1 Software Implementations

The performance of cryptographic software implementations on current desktop personal computers (e.g., a standard 32 bit machine running at 2 GHz) is fairly uncritical by now. As can be seen in Table 3.2, all cryptographic applications using hash functions or symmetric-key primitives normally can be executed without any perceivable delays for a computer user. Moreover, even the execution delays of asymmetric-key operations, which are by orders of magnitude slower than any symmetric-key operation, normally are hardly noticeable for desktop PC users.

Nevertheless, Table 3.3 depicts some general asymmetric-key performance proportions. Thus, a RSA-based signature verification (i.e., based on small public exponents) outperforms both ECC-based signature operations of similar security level, that means, ECC signature generation and ECC signature verification. However, an ECC-based signature generation, in turn, outperforms the RSA-based signature generation of similar security level, whereas ECC signature verifications are by a factor of $2 - 5$ slower than ECC signature generations. Thus, ECC should be used in cryptographic applications on embedded platforms that require both signature generations and signature verifications, whereas cryptographic applications that solely require signature verifications should apply RSA for signature verifications based on small public exponents. Since the sizes for code and RAM, in this context, are comparatively small (i.e., a few kB), they are not mentioned here.

	SHA-1	MD5	3DES	AES-128
Performance	1,000 Mbit/s	1,700 Mbit/s	80 Mbit/s	500 Mbit/s

Table 3.2: Approximate performance values of some selected hash function and symmetric-key software implementations with a similar security level (cf. Table 3.1) on a 32 bit Intel Pentium M processor at 1300 MHz based on the OpenSSL benchmarks [Ope08].

	RSA-1024	RSA-1024	ECDSA-160	ECDSA-160
	Signature	Signature	Signature	Signature
	generation	verification	generation	verification
Performance	50 ms	2 ms	7 ms	31 ms

Table 3.3: Approximate performance values of some selected asymmetric-key software implementations with a similar security level (cf. Table 3.1) on a 32 bit Intel Pentium M processor at 1300 MHz based on the OpenSSL benchmarks [Ope08]. All ECC operations employ SECG-standardized elliptic curves [Sta98].

The performance of cryptographic software implementations on embedded processors (e.g., register widths $\leq$ 32 bit and clock rates $\leq$ 100 MHz), however, can quickly become a critical issue. However, considering the computational context, hash functions and symmetric-key primitives can still be implemented quite adequately regarding cryptographic performance and required memory resources. Table 3.4 also depicts an interesting characteristic of hash functions on small 8 bit processors, which, in contrast to implementations on 32 bit machines or higher, perform considerably worse due to their comparatively large block sizes (e.g., 512 bit). Hence, on very small embedded systems, block ciphers are even able to outperform hash software implementations. Strong asymmetric-key cryptography on small embedded microprocessors, however, becomes challenging. As shown in Table 3.5, asymmetric-key operations on an 8 bit microcontroller can quickly cause execution delays up to a second. Thus, cryptographic applications using asymmetric-key cryptography on small embedded microprocessors have to be designed carefully, but, nevertheless, are practically feasible.

3.7.2 Hardware Implementations

Cryptographic hardware implementations particularly depend on the respective application area. Hence, hardware implementations can be optimized regarding chip area, execution time, or even energy consumption, where each optimization

	SHA-1	MD5	3DES	AES-128
Performance	30 kbit/s	50 kbit/s	10 kbit/s	75 kbit/s
Code size	4 kB	11 kB	4 kB	3 kB

Table 3.4: Approximate values for performance and code size of some selected hash function [GVP+03] and symmetric-key [EKP+07] software implementations with a similar security level (cf. Table 3.1) on an 8 bit microcontroller at 4 MHz.

	RSA-1024	RSA-1024	ECC-160	ECC-192
	Full private key exponentiation	Small public key exponentiation	Point multiplication over GF(p)	Point multiplication over GF(p)
Performance	10,990 ms	430 ms	810 ms	1,240 ms
Code size	6 kB	1 kB	4 kB	4 kB

Table 3.5: Approximate values for performance and code size of some selected asymmetric-key software implementations with a similar security level (cf. Table 3.1) on an 8 bit microcontroller at 8 MHz [GPW+04]. Each of the performance values presents the crucial cryptographic operation for signature generation (i.e., private key exponentiation or point multiplication), signature verification (i.e., public key exponentiation or point multiplication), or for key agreements (i.e., public and private key exponentiation or point multiplication). All ECC operations employ SECG-standardized elliptic curves [Sta98].

has its corresponding trade-offs (e.g., small slow realizations against large fast implementations). Without particular limitations regarding area and energy consumption, dedicated hardware implementations of most cryptographic primitives can reach throughputs of Gbit/s (e.g., for network encryption). In contrast, most embedded-capable hardware implementations have strong restrictions on maximum area consumption (and hence costs), delays, and energy consumption. Due to these mutually conflicting requirements, giving general values is rather difficult. Thus, the following tables only give some rough ideas about performance values and hardware requirements that can be expected.

As shown in Table 3.6 and 3.7, most hash functions and symmetric-key primitives can be implemented in hardware both small and quite fast. In practice, however, the given throughput values (bits/clock cycle) depend on the maximum practical clock frequency, which in turn depends on the underlying semiconductor manufacturing process (e.g., 0.13 μm or 0.35 μm technology) and which inherently affects the actual energy consumption. The area consumption, given in gate equivalents (GE), and hence the costs of the respective digital circuit are moreover rather low. In contrast, hardware implementations of asymmetric-key primitives, as shown in Table 3.8, are considerably more complex, more costly, and require several orders of magnitude more clock cycles. However, at least ECC-based hardware implementations can be realized quite efficiently regarding throughput and area consumption. Moreover, Table 3.8 demonstrates the always possible, different optimizations regarding performance or area consumption (e.g., compare ECC-160 against ECC-163).

	SHA-1	MD5
Performance	512 bit / 1,200 clk	512 bit / 700 clk
Area	8,100 GEs	8,000 GEs

Table 3.6: Approximate values for performance and area consumption of two selected hash function hardware implementations with a similar security level (cf. Table 3.1) using different logic processes [EKP$^+$07, FW07a].

	DES	AES-128
Performance	64 bit / 144 clk	128 bit / 1,032 clk
Area	2,300 GEs	3,400 GEs

Table 3.7: Approximate values for performance and area consumption of two selected symmetric-key hardware implementations with a similar security level (cf. Table 3.1) using different logic processes [EKP$^+$07, FW07a].

	ECC-160	ECC-163
Performance	160 bit / 360 kclk	163 bit / 134 kclk
Area	19,000 GEs	46,000 GEs
	ECC-192	**RSA-1024**
Performance	192 bit / 500 kclk	1024 bit / 2,500 kclk
Area	23,600 GEs	117,600 GEs

Table 3.8: Approximate values for performance and area consumption of some selected asymmetric-key hardware implementations with a similar security level (cf. Table 3.1) using different logic processes [EKP$^+$07, FW07a, Fra08, FW07b].

3.8 Trusted Computing Technology

This section gives a brief overview of the main aspects of Trusted Computing (TC) technology as proposed by the Trusted Computing Group (TCG) [Tru03]. The Trusted Computing technology shall provide the basic security components and functionalities that form the basis for a larger set of security functions that can be built. Together with a secure operating system, Trusted Computing can be used to build an appropriate basis for security architectures with improved security for distributed applications. The main TCG specifications are a component providing cryptographic functions called *Trusted Platform Module* (TPM), a kind of (protected) pre-BIOS (Basic I/O System) called the *Core Root of Trust for Measurement* (CRTM), and the *Trusted Software Stack* (TSS), which is the software interface to provide TC functionalities to the operating system.

TPMs are already integrated into various general-purpose computing devices such as notebooks or desktop computers, but will also emerge into various embedded systems such as cellular phones [Tru05, Tru06]. TPM support is also already integrated into commercial operating systems, e.g., to enable hard disk encryption in Microsoft Vista [Mic08b], or to measure platform integrity under Linux [SS08]. However, the TCG issues only functional specifications while implementations are left to the vendors. Lastly, Trusted Computing technology and systems incorporating TC are not subject to any cryptographic export restrictions [Koo08] and hence can also be employed in IT systems that are applied worldwide.

3.8.1 Trusted Computing Components

As mentioned in the previous section, the most important components specified by the TCG are the *Trusted Platform Module* (TPM), the *Core Root of Trust for Measurement* (CRTM), and the *Trusted Software Stack* (TSS), which all are briefly introduced in the following.

Trusted Platform Module

The base of Trusted Computing technology is the Trusted Platform Module (TPM) that is considered to be a tamper-resistant hardware device similar to a smart-card and is assumed to be securely bound to the computing platform. The TPM is primarily used as a root of trust for integrity measurement and reporting and to secure all critical cryptographic operations. Current TPMs are based on the specification version 1.2 [Tru07b] published by the Trusted Computing Group (TCG), successor of the Trusted Computing Platform Alliance (TCPA), an initiative led by

AMD, HP, IBM, Infineon, Intel, Lenovo, Microsoft, and Sun. TPMs are available, for instance, from Atmel, Broadcom, Infineon, Sinosun, STMicroelectronics, and Winbond.

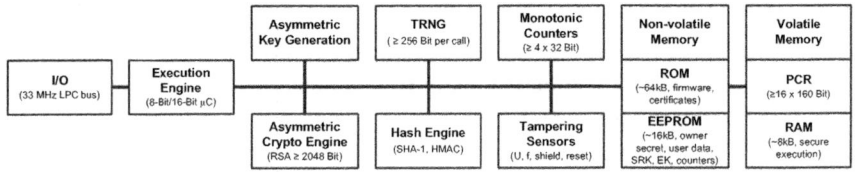

Figure 3.8: Internal structure of a Trusted Platform Module (TPM) according to the recent TCG specification version 1.2 [Tru07b].

Figure 3.8 illustrates the components of a TPM according to the most recent specification version 1.2 [Tru07b]. A TPM provides the following features: A hardware-based random number generator (RNG), a cryptographic engine for encryption and signing (RSA) as well as a cryptographic hash function (SHA-1, HMAC), read-only memory (ROM) for firmware and certificates, volatile memory (RAM), non-volatile memory (EEPROM) for internal keys, monotonic counter values and authorization secrets, and, optionally, sensors for tampering detection. Common TPM chips use a synchronous Low Pin Count-I/O-Interface (LPC-I/O) to communicate with their host system (underlying processor). Based on protected hardware functionality the security-critical operations and information like key generation and decryption are performed on-chip. Secret keys never leave the device unencrypted.[7] A TPM can be abstractly described by the tuple $\mathcal{T} = (EK, SRK, TD)$ with the endorsement key EK, an asymmetric key that uniquely identifies each TPM; the Storage Root Key SRK, an asymmetric key used to encrypt all other keys created by the TPM; and the TPM data TD that includes further security-critical non-volatile data shielded by the TPM. Note that neither EK nor SRK can be readout from the TPM. The TPM provides a set of registers called *Platform Configuration Registers* (PCR) that can be used to store hash values. The hardware ensures that the value of a PCR can only be modified as follows: $PCR_{i+1} \leftarrow \mathsf{Hash}(PCR_i|x)$, with the old register value PCR_i, the new register value PCR_{i+1}, and the input x (e.g. a SHA-1 hash value). This process is called *extending* a PCR.

---

[7]To perform a decryption operation with a specific key, several types of authorization are possible.

Core Root of Trust Measurement

The Core Root of Trust Measurement (CRTM) represents immutable code implemented into the boot ROM that is executed at first during the booting process to initialize the root of trust of the corresponding device. Therefore, the CRTM initializes a hierarchical hash chain starting with itself followed by a step by step hashing of the program code of the respective upper layers such as the remainders of the boot ROM, the boot routine, the boot loader, and all consecutive layers (cf. Figure 3.9). Thus, since the security of the chain explicitly relies on the security of the CRTM, the CRTM has to be trusted a priori by all involved parties.

Trusted Software Stack

The TCG Software Stack (TSS) [Tru07a] is comprised of modules and components that present the software interface (API) to provide the functionality of the TPM to the operating system. The TSS provides the TPM device driver and presents common cryptographic interfaces to require only minimal modifications on existing applications. It handles internal keys management, the command synchronization and controls all critical machine-level invocations. As the TSS handles security-critical data and operations, it has to be a trusted OS component.

3.8.2 Trusted Computing Basic Functionalities

Based on the TC components specified by the TCG, Trusted Computing technology provides some basic security functionalities on which a larger set of security mechanisms can be built [Tru07a, Tru07b]. Hence, these base security functionalities are as follows.

Secured Cryptography

Trusted Computing hardware implements a set of cryptographic operations to ensure that malicious software cannot compromise cryptographic keys. Usually, key generation and decryption operations are done "on-chip". Private and secret keys never leave the chip without being encrypted. To perform a decryption operation with a specific key, several types of authorization are possible. A distinctive feature of Trusted Computing hardware is the ability to not only use passwords as authorization but also integrity measurements. That is, only a platform running previously defined software or hardware components is authorized to use a certain key. Moreover, the property that a certain key is "bound" to a platform configuration can be certified by Trusted Computing hardware. This certification includes the integrity measurements which authorize a platform to employ the key. A remote party can verify the certificate and validate the embedded integrity measurement against "known good" configurations before encrypting data with the certified key[8].

Authenticated Booting

During an authenticated boot process, each part of the code which is executed is "measured" before execution, for instance, by calculating a cryptographic hash of the code. Trusted Computing hardware, in turn, is responsible for the secure storage and provision of the measurement results.

For authenticated boot, the PCR values are used to establish the *chain of trust*, which is exemplarily shown in Figure 3.9. Therefore, at power-on of the platform, the CRTM computes a hash value of the code and parameters of the BIOS that includes the corresponding hardware controlled by the BIOS. Computing the hash value is called *integrity measurement* in the TCG terminology. Then the CRTM extends the corresponding PCR and hands over the control to the BIOS that measures the next component, for example, the boot loader, and again extends the corresponding PCR before switching the control to it, and so forth. The security of the chain relies strongly on explicit security assumptions about the CRTM. Upon completion of an authenticated boot process, these measurement results reflect the hardware and software configuration of the computing device. Thus, the set of PCR values $PCR_0, \ldots, PCR_n$ provides evidence of the system's state after boot. We call this state the platform's *configuration*, denoted by $C := (PCR_0, \ldots, PCR_n)$. Trusted Computing technology, however, remains passive and does explicitly not prevent a certain computing environment from being executed or compromised during runtime. The TPM uses the measured configuration as base for most of the following security mechanisms.

Binding and Sealing

The TC binding mechanism enables to cryptographically bind arbitrary data to a certain device hardware and software configuration C and/or device identifier. For this, the TPM internally creates a certified asymmetric key pair, where the private part is protected by the TPM. The corresponding public key is further certified by the TPM, where the certificate states that the configuration C_i is required to access and to use the protected private key. Hence, the usage of the private key is "bound"

---

[8]Mostly, the data to be encrypted is a secret symmetric key used for the encryption of larger amounts of data.

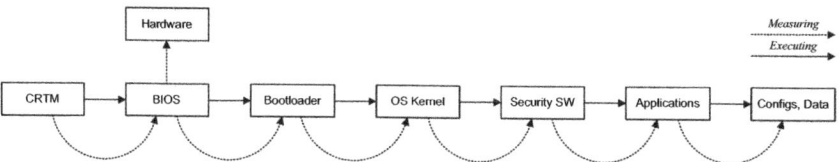

Figure 3.9: Trusted Computing instance of the "chain of trust for measurements" where the root of trust resides at the CRTM. Hence, starting at the CRTM, each component *first measures* the next component *before executing* it and hence handing control over to the subsequent component.

to the corresponding device identifier and/or configuration C that is reflected by a subset of PCRs. Since the certification includes the parameters which authorize a device to employ the key, a party can verify the certificate and validate the embedded parameters against "known good" reference configurations before encrypting data with the certified key.

The TC sealing mechanism is an analogous mechanism to binding, however with mandatory linkage to the device identifier based on TPM's individual *EK*. Thus, sealing is mainly used for sealed storage, i.e., to cryptographically bind the data of the device to itself.

Remote Attestation

The remote attestation is used to give assurance about the platform configuration C to a remote party. To guarantee integrity and freshness, this value and a fresh nonce provided by the remote party are digitally signed with an asymmetric key called *Attestation Identity Key* (AIK) that is linked to the *EK* and is under the sole control of the TPM. A trusted third party called *Privacy Certification Authority* (Privacy CA) is used to guarantee the pseudonymity of the AIKs. In order to overcome the problem that this party can link transactions to a certain platform, version 1.2 of the TCG specification defines a cryptographic protocol called *Direct Anonymous Attestation* based on a zero-knowledge proof [BCC04], eliminating this CA.

Trusted Channels

A trusted channel is a secure channel[9] that is additionally bound to the configuration C of the endpoint. More concretely, the trusted channel additionally allows

---

[9]A secure channel ensures confidentiality and integrity of the communicated data as well as the authenticity of communication endpoints.

each endpoint component to (i) validate the configuration of the other endpoint component and (ii) to bind data to the configuration of the endpoint component such that solely and exclusively this component with this configuration can access the data. Trusted channels can be implemented either by directly employing TC technology [AES$^+$07, SSSW06], or by using a more generalized approach that is also able to employ other hardware security architectures [AOS$^+$08].

3.9 Security Schemes in the Automotive Domain

The cryptographic functionalities used for many automotive applications are *digital signatures*, *hybrid encryption*, and the *challenge-response* protocol.

3.9.1 Digital Signatures

Digital signatures are the digital counterpart to handwritten signatures. Analogous to hand-written signatures on a paper, digital signatures enable the authentication of a digital information, that means, they refer to certain information *and* a certain party. They further can ensure information integrity and non-repudiation. A digital signature usually involves two algorithms, (i) an asymmetric-key algorithm, which generates the keys and provides the corresponding schemes for signature generation and signature verification, and (ii) a cryptographic hash function, which reduces an information of any length to a (nearly) unique "digital fingerprint" of a fixed length that is a concise representation of the longer information. Thus, the basic steps involved in generating and verifying a digital signature s for an information m are as follows.

(1) An asymmetric-key algorithm generates an asymmetric key pair, namely the private key SK_{SIG} for signature generation and the corresponding public key PK_{SIG} for signature verification. The *signer* keeps SK_{SIG} secret, whereas PK_{SIG} becomes publicly available (or at least for all *verifiers*).

(2) The *signer* uses a cryptographic hash function Hash to compress the information m to its hash value h_m of fixed length such that $h_m = \text{Hash}(m)$.

(3) The *signer* uses the asymmetric-key algorithm and the secret signing key SK_{SIG} to produce the digital signature s such that $s = \text{Sign}(h_m, SK_{SIG})$.

(4) The *verifier* of a signed information receives m, s, and PK_{SIG}.

(5) The *verifier* individually computes the hash value $h_m = \text{Hash}(m)$.

(6) The *verifier* uses the asymmetric-key algorithm together with the public verification key PK_{SIG} and the individually derived fingerprint h_m to accept or reject the corresponding digital signature s via $\text{Verify}(h_m, s, PK_{SIG})$.

Thus, a digital signature provides all important properties of a handwritten signature, namely:

- Only the holder of the secret signature generation key PK_{SIG} is able to sign an information properly.

- Anyone can verify the signature by using the corresponding public signature verification key PK_{SIG}.

- The signed information and signature are directly related, that means, a (valid) signature cannot be reused to sign another information or a subsequently changed information.

Digital signatures are used already in many applications such as electronic mails or electronic banking to provide information integrity, information authenticity, and non-repudiation. Note that digital signatures provide security properties (e.g., non-repudiation), which cannot be provided by applying symmetric-key algorithms howsoever. Since symmetric-key algorithms are based on a shared secret, *each* party that has access to the shared secret (i.e., the signer and the verifier) would be able to create valid signatures. Thus, symmetric message authentication codes, for instance, cannot protect two parties against each other.

However, digital signatures certainly cannot provide all security properties. For example, they do not inherently provide information confidentiality and any freshness detection (e.g., assurance about date and time at which an information was signed). They further often rely on costly public key infrastructures (cf. Section 3.9.3). Nonetheless, for information confidentiality, the information can be additionally encrypted either by using an additional shared secret or the public key of the receiver, whereas freshness detection can be achieved by proper time stamping.

3.9.2 Key Exchange and Hybrid Encryption

A major disadvantage of asymmetric-key primitives, when compared with symmetric-key schemes, are the intensive arithmetic operations that need to be performed. Hence, this can lead to a poor overall system performance. Even when properly implemented, all asymmetric-key schemes proposed to date are several

orders of magnitude slower than the most efficient symmetric-key schemes (cf.
Section 3.7). Hence, in practice, cryptographic systems are applied as a mixture of
symmetric-key and asymmetric-key cryptography in a hybrid fashion. Therefore,
an asymmetric-key algorithm is applied for exchanging a secret key and then a
symmetric-key algorithm is applied to encrypt the communication of bulk data to
achieve high throughput values.

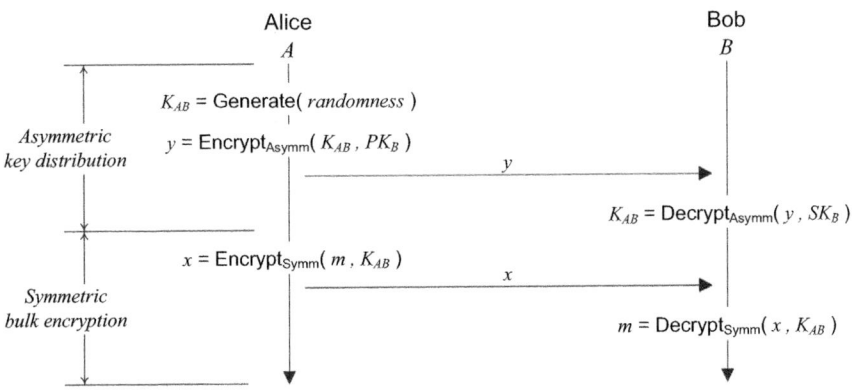

Figure 3.10: Exemplary hybrid encryption protocol where Alice first generates and distributes a secret key K_{AB} using an asymmetric-key algorithm Asymm followed by the
mutual bulk data encryption using a symmetric-key algorithm Symm and the previously
exchanged secret key K_{AB}.

Figure 3.10 shows an asymmetric *key distribution* protocol, where one party
(Alice) generates a secret key and distributes the key to the other party (Bob).
Thus, Alice first encrypts a secret symmetric key K_{AB} with the public key PK_{Bob} of
Bob. Bob then decrypts K_{AB} using his secret private key SK_{Bob}. Afterwards, both
proceed their communication using a symmetric-key algorithm with the secret key
K_{AB} exchanged before. Another, more sophisticated asymmetric *key agreement*
protocol, where the involved parties jointly generate a secret key, is the widely
used Diffie-Hellman Key Exchange (DHKE) [DH76].

If only symmetric-key cryptography is available, the shared secret key can be
securely established using a trusted third party (TTP) or a so-called key distribution center (KDC) [NHR05]. However, for the secure communication with the
TTP/KDC, the involved parties again need either an a-priori secret (initial key) or
a physical secure channel.

3.9.3 Public-Key Infrastructure

By applying asymmetric-key algorithms for hybrid encryptions or signature verifications, the encrypting or verifying party has to rely on the authenticity of the respective public key. This means, if a malicious entity could replace the proper public key PK by foisting a public key PK_M, where the malicious entity has access to the corresponding private key SK_M, the malicious entity could decrypt all encrypted information or foist self-generated signatures. To prevent such "man-in-the-middle" attacks as shown in Figure 3.11, the user of a public key has to be sure that the applied public key comes in fact from the entity he expects. However, requiring an additional communication channel to securely receive the respective public keys would render most asymmetric-key schemes worthless.

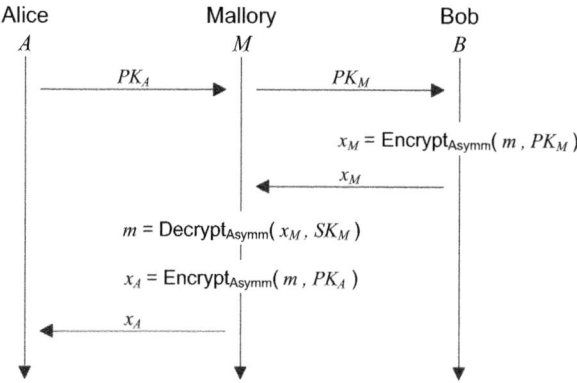

Figure 3.11: Exemplary *man-in-the-middle* attack on an asymmetric-key encryption where the malicious entity M replaces the original public key of the receiver PK_A with its own public key PK_M. Thus, M can decrypt all information from the sender B while re-encrypting all information for the intended receiver A to remain unnoticed by both.

A public key infrastructure (PKI), however, can provide authentication of public keys by cryptographically linking a public key to the respective identity (e.g., a person or an institution). To accomplish that, public key infrastructures employ a trusted third party known as certificate authority (CA), which, depending on the level of assurance, verifies the proper linkage of a certain identity and the respective public key (e.g., by a registration process under human supervision). After a successful verification, the CA issues a public key certificate (or identity certificate), which cryptographically links the public key and the corresponding

identity by means of digital signatures (cf. Section 3.9.1). Thus, a public key
certificate *cert* contains the respective public key PK, the corresponding identity
ID, and a digital signature SIG_{CA} of the certificate authority over both as follows.

$$cert = \{PK, ID, SIG_{CA}\} \quad \text{with} \quad SIG_{CA} = \text{Sign}[\text{Hash}(PK, ID), SK_{CA}]$$

Beyond PK and ID, a certificate could further include certain validity conditions and other attributes in an analogous manner. However, to verify that a certain
public key belongs to a certain entity, the verifier applies the public key PK_{CA} of
the (trusted) certificate authority on the corresponding certificate and checks the
proper linkage between the public key and the identity by means of a digital signature verification. In order to avoid that a verifier has to (securely) store a large
number of public keys of trusted CAs, CAs are often hierarchically structured such
that so-called root-CAs issue certificates for the public keys of the lower-level CAs
and so on, that results in a tree structure of hierarchically verified certificates. In
order to thwart changed identities or compromised private keys, certificates usually
have only a limited period of validity or can be added to so-called certificate revocation lists (CRL), which are checked during the certificate verification process.
Whereas, comprised or malicious CAs can be thwarted by the proper application
of threshold cryptography [Des94] or methods of secret sharing [Sha79].

Lastly, a PKI comprises all hardware and software, the personnel, and all principles and practices involved for certificate generation, certificate administration,
certificate distribution, and certificate revocation. The currently most popular and
widely used standard for public key infrastructures, certificates, and certificate revocation lists is the internet X.509 standard [HFPS02].

3.9.4 Challenge-Response Protocol

A challenge-response protocol provides entity authentication, also called identification, that is, one communication party identifies itself to a second party. The
identification can be provided by using knowledge, possession or individual properties. The basic idea of the challenge-response protocol is that one party challenges the second party, for instance, by sending a random number. The chal-
lenged party then has to answer with the correct response. This correct response
can be generated only if the second party has some kind of knowledge, for example, the key for a cryptographic primitive. The party can use the key to encrypt
the given random number and returns it to the challenger, thus proving the possession of knowledge without revealing it. The protocol can be implemented using
symmetric-key as well as asymmetric-key primitives.

Figure 3.12 presents a challenge-response protocol using a symmetric-key algorithm Symm. In this scenario, Alice and Bob share a secret key K_{AB}. Therefore, Alice challenges Bob by sending a random number c_A that she has created before. Bob encrypts c_A together with the identity ID_A of Alice to r and returns the response r to Alice. Bob is identified once Alice has decrypted r and could properly verify the resulting identity ID_A and the random challenge c_A. Note that only Bob can response the challenge correctly, since (next to Alice) only Bob possesses the knowledge of the appropriate secret key K_{AB}.

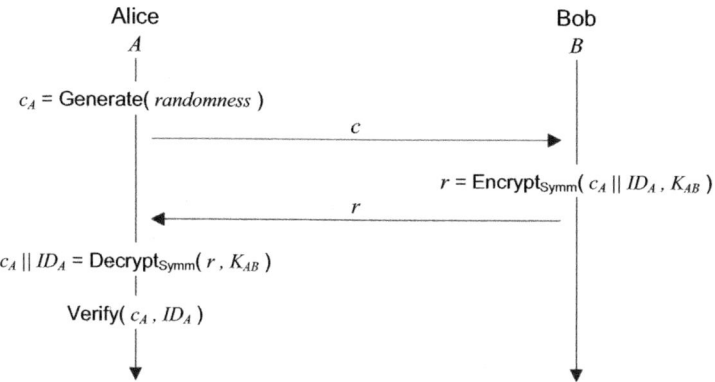

Figure 3.12: Exemplary challenge-response protocol, where Alice generates a random challenge c_A used to challenge Bob, who replies with a proper encryption of c and Alice's identity ID_A using the symmetric-key algorithm Symm. After decryption of r and successful verification of c and ID_A, Bob is successfully identified to Alice.

Examples of more sophisticated challenge-response protocols are *zero-knowledge password proofs* and *password-authenticated key agreements*, which both can be found in [IEE00].

3.10 Cryptanalysis

As mentioned in the introductory section of this chapter, cryptanalysis is next to cryptography the other integral part of cryptology. Cryptanalysis comprises all methods for obtaining protected information without having access to the corresponding secrets and thus is inevitable to continuously evaluate and assure the security of cryptographic schemes and IT security mechanisms in general. As shown in Figure 3.1, cryptanalysis therefore employs not only classical *mathematical attacks*, but also attacks on *implementational weaknesses* or methods of *social engineering*. However, despite popular beliefs about codebreaking, most security mechanisms were broken without applying any mathematical attacks on the underlying cryptography by exploiting weaknesses in the internal structures (i.e., *analytical attacks* such as linear and differential cryptanalysis) or trying all possible keys (i.e., *brute-force attacks*). Thus, as more detailed described in Section 5.2.3, attacks on organizational weaknesses or simple social engineering attacks can often yield much faster and less costly to the disclosure of valueable secrets than laborious mathematical attacks (if computationally feasible at all).

Another important area of cryptanalysis are attacks on the actual implementation of cryptographic methods. This means for instance various kinds of physical attacks such as side-channel attacks or fault attacks (cf. Section 5.2.2), or software attacks that exploit insecure software interfaces and (unintended) functionality using corrupted inputs or buffer overflows (cf. Section 5.2.1). However, cryptanalysis is a science on its own and hence further readings on this subject are helpful and can be found amongst others in [Fri93, LR07, Sch06, Sin98].

Lastly, a potential attacker will unlikely attack the underlying cryptography, if some organizational vulnerability turns out to be the lower hurdle to breach the security of the particular IT system. That means, the overall security of a vehicular IT system is not defined by the strength of its individual security measures, but solely by the security of its weakest link.

Part II

The Threats

4 Security-Critical Vehicular Applications

This chapter identifies various security-critical vehicular applications. This includes vehicular mechanisms and components, IT-based business models, legal applications as well as communication and safety applications that are current state-of-the-art, but also includes several future security-critical vehicular applications. Parts of this chapter are based on published research in [BEPW07, LPW06, WWW07].

4.1 Introduction

Security-critical automotive applications refer to applications that require—additionally to measures against random technical failures (i.e., IT safety measures)—measures for the protection against malicious manipulations. Thus, security-critical automotive applications are applications that somehow require prevention of unauthorized access, modification, or destruction of their data or functionality. So far, there only existed niche applications in the automotive domain (e.g., immobilizers) that particularly rely on IT security technologies. However, the situation has changed dramatically. More and more vehicular systems need security functionalities in order to provide driving and road safety as well as to protect the revenues, liability, and reputation of manufacturers, component suppliers and the (economic) interests of the vehicle owner. In the following, an overview about current and future security-critical vehicular applications is given.

4.2 Theft Protection

Anti-theft applications are probably the oldest security-critical automotive applications and thus probably induced the first application of cryptography in vehicles. They basically are supposed to prevent the theft of a vehicle and hence handle the authentication of the drivers authorized to use the vehicle. The two most important applications in this area are the *electronic immobilizer* and the *electronic key* as described in the following.

4.2.1 Electronic Immobilizer

An electronic immobilizer should prevent the vehicle's engine from starting in case a thief already managed it to break into the interior of a vehicle. It is already legally mandatory for all new vehicles in many countries (e.g., Germany, United Kingdom, or Australia) and obligatory for most car insurance policies.

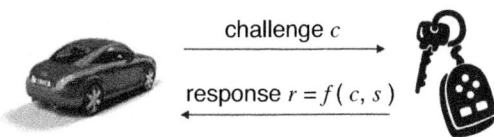

Figure 4.1: Schedule of a challenge-response protocol used in an electronic immobilizer.

An electronic immobilizer is automatically activated after vehicle shutdown and prevents the vehicle's starting system from starting the engine again, unless the correct authorization is provided. This authorization is mostly based on a passive battery-less RFID[1] transponder integrated in the vehicle key and the corresponding RFID reader integrated in the starting system of the vehicle. Electronic immobilizers usually employ a challenge-response protocol as described in Section 3.9.4. Therefore, as depicted in Figure 4.1, the vehicle first sends a challenge c to the passive RFID transponder integrated in the vehicle key. This can be done for instance upon driver's request, when he tries to enter the corresponding vehicle. The key transponder then answers with a response r based on the respective challenge c and the secret s shared between transponder and vehicle without revealing it. Here, f is a cryptographic function such as a keyed hash function that takes as input the challenge c as well as a secret s and returns the corresponding response r, that could be for example the hash value of c and s. Hence, a potential car thief could neither replay an afore intercepted authorization message, nor gain any further knowledge about the shared secret s by eavesdropping the respective authentication sessions. Further, in contrast to a solely mechanical key, an exact physical copy of an electronic vehicle key, which does not has the cryptographic secret s, will fail to provide a valid response r and hence fail to start the vehicle.

Breaking correctly implemented electronic immobilizer usually requires attacks at the hardware layer. A potential attacker could for instance try to read out and copy the secret s from the corresponding vehicle key. However, here the attacker must have access to the original key. Otherwise, an attacker could replace the start-

---

[1]Radio Frequency Identification [HC06]

ing system of the vehicle completely by one that is always activated, for example by one that implements exactly the same functionality but without the respective verification procedure. Such hardware attacks can never be prevented at reasonable cost. However, the goal is to make such an attack infeasible for a rational attacker, that means, the cost of an attack shall exceed the potential gain of the stolen vehicle on the black market. This is also known as economic security. Furthermore, a so-called relay attack (or Mafia attack or man-in-the-middle attack) is possible.

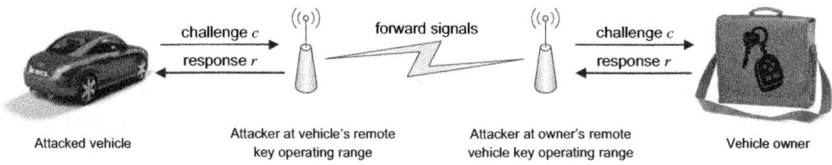

Figure 4.2: Scheme of a relay attack where an attacker intercepts and forwards the authentication procedure from the attacked vehicle to original vehicle key unnoticed by actual vehicle owner.

As shown if Figure 4.2, the vehicle's challenge is (covertly) forwarded over an extra channel (say, an intermediate wireless connection) to the original vehicle key, which then automatically computes the correct response that in turn is sent back to the corresponding vehicle to deactivate the electronic immobilizer. The latter attack can be thwarted for instance by demanding an acknowledgment (e.g., pushing of a button) first, before the vehicle key returns the corresponding response, by the additional application of so-called distance bounding mechanisms [CH06], or by applying a specific communication technique between key and vehicle, which is very difficult to forward (e.g., frequency hopping or signal modulation). Further reading on a general approach for implementing a secure electronic immobilizer can be found for instance in [LSS05].

4.2.2 Remote Door Lock

A remote (keyless) door lock system automatically locks or unlocks the doors of a vehicle without the need to physically put the key in the door lock cylinder of the vehicle. The remote door lock works in a similar way as an electronic immobilizer and hence can also include the authentication to the engine starting system to enable for instance remote keyless ignition. In a remote door lock system, the vehicle key is equipped with an active battery-powered transponder that sends the respective lock or unlock message m when pushing the corresponding button on the vehicle key, or automatically if, the corresponding vehicle key token enters or leaves the operating range of a keyless door lock system. For that, so-called rolling codes can be used, where vehicle and vehicle key share a secret s and synchronize a monotonic counter i.

As depicted in Figure 4.3, to provide a valid lock or unlock message m, the vehicle key first increments its monotonic counter i_K and invokes a cryptographic function f together with i_K and s. Here, f could again be a keyed hash function or an efficient block cipher[2]. After the vehicular key has sent m to the vehicle, the vehicle computes the same transition and compares its own result with the received message m to validate the authenticity of the message before executing the corresponding lock or unlock command. To thwart the case that the vehicle accidentally missed some messages m, the vehicle usually compares also the next 256 transitions before demanding further authentication measures. Additionally, some keyless door lock systems also apply a challenge-response protocol as described for electronic immobilizers in the previous section.

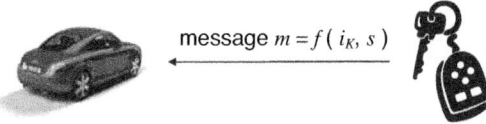

message $m = f(\,i_K, s\,)$

Figure 4.3: Schedule of a rolling code used in a remote door lock system.

Attacks on door look systems are rather hardware attacks — an adversary could otherwise just smash a window to get into the vehicle. Attacks trying to replay a used message or attacks trying to compromise the shared secret can be securely thwarted through the cryptographic scheme. However, attacks on the actual physical transmission of the signals such as a relay attack (cf. Section 4.2.1) or jamming attacks, which try to jam the transmission channel between key and vehicle while intercepting a valid m for later misuse ("code grabbing"), have to be considered carefully since they are inherent for any such scheme and can hardly be prevented by cryptographic means.

---

[2] A widely used algorithm for generating rolling codes in the automotive domain is KeeLoq [Mic08a].

4.3 Counterfeit and Intellectual Property Protection

Today large amounts of OEM's capital investments are spent on software and electronic development [Bro06, SW03] that—without further protection—can be copied easily, analyzed and possibly exploited just by buying the corresponding components or vehicles available on the free market. Worldwide counterfeit and piracy losses are about $3 trillion dollar (USD) per month [Gie05]. Moreover, illegally produced vehicular spare parts cause legitimate manufacturers a worldwide loss of about $12 billion per year [Gar03]. Solely the Ford Motor Company reports losses due to fake parts of about $1 billion a year [CNN07]. Thus, reliable counterfeit, company secrets, and intellectual property (IP) protection represents a very important application area for IT security in vehicles in order to prevent copyright infringement, theft of trade secrets, or expertise theft by potential competitors and particularly to prevent the illegal mass production of counterfeits of vehicle components.

4.3.1 Counterfeit Protection

Professionally organized manipulation of vehicular electronics [And98] causes considerable damage to the manufacturers and to the economics by unwarranted claims and undermined business models. Moreover, counterfeits endanger the safety of all motorists and cyclists and thus strongly damage the public confidence in the brands concerned. Traditional methods to prevent counterfeits use tags, e.g., holographic stickers that are supposed to be unforgeable. However, there exist illegal businesses that create boxes, labels and other significant trademark logos and emblems to let counterfeits look like real parts [Ros04]. Hence reliable counterfeit protection of digital components requires the application of IT security measures to prevent unauthorized access to crucial internals.

4.3.2 Intellectual Property Protection

Automotive OEMs and suppliers always have a comprehensible interest to find out valuable expertise or trade secrets from their competitors. On the other hand, automotive OEMs devote considerable financial and personal resources into the development of vehicular electronics and software to maintain or extend their current market position [Bro06, Fri04]. Even though intellectual property (IP) rights are legally effective in most countries in the world, there exist large domestic markets where IP thefts and copyright infringements are virtually non-triable, if they can be proven to at all [Spe06]. Trade secrets, in contrast to patents, by definition do not have any formal protection, that means a third party is allowed to use a trade or company secret once it has been discovered for instance by lawful methods of reverse engineering or by less lawful methods of industrial espionage.

Thus, expertise leakage, reverse engineering, IP thefts, and copyright infringements are serious issues also in the automotive domain. Today, this applies mostly for software and firmware, but even complex vehicular hardware will be copied when it is profitable enough. Expertise leakage, trade secret disclosure, and IP theft has to be tackled primarily by applying organizational security measures such as scrutinizing potential partners and preventing employees from unintentional (or intentional) exposures (cf. Chapter 9). However, it also requires protection measures inside the actual vehicle in order to prevent reverse engineering and infringements or making IP thefts for instance at least detectable or provable.

4.4 Software Updates

As already mentioned in the introductory Section 1.1, vehicular components are more and more based on flexible, easily customizable software than on specially adapted hardware. Software driven automotive components enable considerable cost reductions (e.g., due to code reuse, easy copying, large-scale application of standard hardware), weight reduction and hence less fuel consumption, and, in particular, enable sophisticated, novel and variable functionality that is hardly feasible solely in hardware. There are estimates that in 2015 over 35% of the total development costs of a new vehicle are related to vehicular software development [Bro06, SW03], while at the same time, software is virtually the driving force behind all vehicular innovations [Fri04].

As depicted in Figure 4.4, a premium vehicle of today has up to several hundred megabytes of embedded code providing more than 2000 individual functions [Bro06] distributed over up to 80 individual ECUs. Future vehicles will include more than one gigabyte of embedded software, which then probably employ only up to three or five centralized powerful controllers [CCA02, Fri04]. It is hardly possible to maintain such an amount of code without requiring any corrections during the comparatively long life cycle of a typical vehicle. Thus, software updates or upgrades in order to fix faults found, to improve or enhance available functionality, or to adapt existing functionality to new regulations will become an everyday routine. Applications for ECU software updates range from occasional bug fixing of software errors in a garage, up to on-the-fly software updates during a fuel stop to receive for instance individual ABS/ESP control parameters specially adapted to the respective tires mounted.

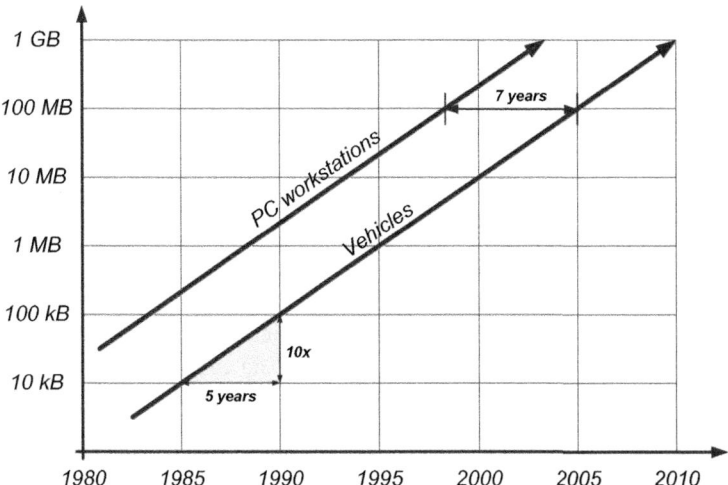

Figure 4.4: Memory usage in vehicles grows likewise exponentially according to Moore's Law [Moo65] and is only about seven years behind the memory usage of general-purpose desktop computers [Fri04].

However, enabling ECU software updates also enables the feasibility of (malicious) ECU software manipulations, unauthorized software substitutions and eases software thefts. Enabling software updates without any restraints hence could undermine many after-sale and legal applications (cf. Section 4.5 and Section 4.6), since an unrestricted software update would enable arbitrary software manipulations. Without further measures, it is further relatively difficult to provide an evidence that a vehicular software is or was manipulated in case of liability or warranty disputes. Thus, a reliable software update usually has the following security objectives.

(1) Only authorized[3] software can be installed in the respective vehicle,

(2) all software changes are recorded in a mutually tamper-proof location, and

(3) integrity and confidentiality of update data can be securely enforced.

---

[3]For liability reasons, vehicular software can usually only be authorized by the corresponding component manufacturer. Since unauthorized software updates also affect various safety issues (e.g., software interoperability), authorization of software updates is obviously not only a security issue.

This requires for instance efficient security measures that enable mutual authentication, secure communication as well as secure storage while enforcing confidentiality, integrity, authenticity, and non-repudiation for the corresponding software update.

4.5 After-Sale Applications

Embedding IT security in vehicles enables various new after-sale business models, which were not possible in this way before. Particularly, it enables business models where all involved parties (OEMs, suppliers and customers) can benefit from. In the following, some new exemplary business models, which would become feasible by the proper application of advanced vehicular IT security technologies.

4.5.1 Feature Activation

To meet the strictly increasing cost pressure, the production manner of vehicular components changes from small charges of different, individually adjusted components towards a large-scale production of only a small variety of uniform standard components. On the other hand, providing manifold individual vehicle configurations is more and more crucial to be successful on nowadays very diversified vehicle market. To solve these opposing requirements, vehicle manufacturers could build parts, identical in physical construction, very cost-efficiently with most features already built-in, but individually activated. Thus, today many of the apparently individual vehicle configurations are internally mostly made up from physically identical components. For instance, a freely configurable dash board display would display only sensors and functions actually available or activated. Moreover, it is possible to individually activate already built-in hardware components or software after the actual sale of the vehicle for an additional charge. Such an after-sale feature activation would enable lower basis prices for the vehicles and would furthermore forge close and long-term links between customers and the corresponding OEM. Features that would be capable for an automotive after-sale activation could be for instance: special setups for engine, gear, or chassis control, enhanced board computer and comfort diagnosis functions, additional driving assistance and infotainment capabilities or certain personalization and individualization features. However, capable security measures are required to prevent unauthorized feature activation that for instance may undermine the underlying business model or that even may compromise the safety of a vehicle.

4.5.2 Infotainment

Probably, some of the most promising applications in the automotive area are driven by new infotainment business models distributing digital content for in-vehicle usage. The area ranges from individual software upgrades, OEM premium content to newscasts including the distribution of arbitrary multimedia files such as music, videos, or games. Today, already most medium-sized vehicles are equipped with multimedia capable on-board computers and radio systems. Upcoming integrated wireless broadband communication promises a brisk market for automotive related on demand aftermarket sales. Embedding a reliable digital rights management (DRM) enables various business models for usage-metered and on-demand utilization of digital contents, software and even hardware beyond the classical lump-sum models. Some possible examples are provided below.

Time-limited utilization. Up-to-date navigation and traffic data may be available on demand for any place in the world but only as long as needed (e.g., buying navigation data for a vacation trip only for the period and for the region that is actually needed).

Quantity-limited utilization. Movies, music tracks, or games could be bought such that their usage is somehow quantitatively limited (e.g., providing n free trials for a game or, instead of selling a movie as a whole, just selling the (cheaper) right for two views).

Device-bound utilization. The execution or installation of certain software can be restricted to devices or vehicles of a particular type or brand, or is possible only if a certain combination of other devices is available. Some vehicular functions could be for instance available only if the corresponding authentication device such as the driver's key, a dealer's token, or the personal cellular is available.

Usage-metered utilization. Navigation routes can be charged for their length actually used. Movies or music tracks can be charged for the actual viewing/listing time. Premium OEM information could be freely available for a certain amount of kilometers traveled.

Subscription services. Audio, video or information broadcast services can be received as long as a valid subscription to the corresponding service exists.

Furthermore, almost arbitrary combinations are possible. For instance, an afterwards activated enhanced comfort sensor (e.g., a tire air pressure sensor) could be enabled for four weeks as free sample. As usual in various (non-automotive) multimedia scenarios that employ digital content that has usage or access restrictions such as pay-TV, online music stores, or video game consoles, the usage of protected infotainment content in the automotive domain requires similar protection measures against misuse or circumvention of existing restrictions.

4.6 Legal Applications

Supporting new vehicular legal applications could become a crucial impulse for automotive electronics. Since official applications normally have great financial, legal, and personal meanings, they require strong IT security measures to prevent (or at least detect) any kind of manipulations. Moreover, as official applications also often involve many sensitive personal information, they furthermore require strong IT security measures for privacy protection. In the following, some current and future legal vehicular applications, which require IT security measures, are presented.

4.6.1 Milage Counter

The vehicular milage counter is probably one of the eldest targets for systematic manipulations and, along with the introduction of electronic milage counters, one of the eldest application scenarios of vehicular IT security. The attacker, who is usually also the actual owner of the vehicle, tries to decrease the actual milage in order to achieve better prices on the resale of the vehicle or to undermine existing warranty programs or leasing contracts. In contrast to this, attackers might also try to increase milage of the vehicle to obtain for instance a higher tax refund or to fiddle expenses. Today, it is assumed[4] that about a third of all used cars have manipulated milage counters.

Particularly, the introduction of electronic milage counters made malicious manipulations very easy and hardly detectable. In fact, even though such manipulations are generally illegal, there exists nearly a regular business that provides arbitrary milage manipulations for almost every vehicle for about 100 dollars just in a few minutes. Due to the potential millionfold resale of a once-successful attack method, attackers are able and willing to invest large financial, personal, and technical resources for breaking a new milage protection mechanism. Attackers therefore can try to manipulate the respective motion sensor, the storage location(s) of the milage value, or the communication link between milage storage and motion sensor. For luxury vehicles, it could even be profitable to simply exchange the complete dashboard module.

---

[4]This is an unofficial estimation of the German DEKRA [Deu08], one of the world largest motor vehicle assessor organizations.

Since the attacker usually has full physical access to the vehicle and can reuse expert knowledge gained afore, it is difficult to achieve even economic security such that the total cost of a successful attack exceeds its potential economic gain. Thus, a proper milage counter protection requires additional organizational measures (e.g., a central vehicle data base) and sophisticated IT security measures using high redundancy mechanisms, which then require costly individual attacks each time, and whose results cannot be reused in any further attacks.

4.6.2 Electronic License Plate

Integrating a wireless transponder into a vehicle that broadcasts a unique identification string will be an important automotive application in the near future that makes the error-prone and easy manipulable metallic vehicle license plate obsolete. In fact, first implementations are already available [e-P08] and they were already successfully tested in a feasibility trial by the Department for Transport in the United Kingdom in 2006 [DD06]. The Road Transport Department of Malaysia already in 2007 starts to implement microchip-enabled licence plates in all vehicles to stop thieves from stealing such specially marked vehicles [Kau06].

Such an electronic license plate could furthermore help to easily implement wireless tolling and payment systems or simplify the return of a rental car. It could particularly help police forces and public authorities to easily identify a vehicle in case of accident or law violation and to impede any illegal change of the license plate as well as vehicle theft. On the other hand, an electronic license plate enables, in comparison to today's always fully visible metallic plates, a privacy-enhancing fine-grained access control policy up to complete access refusal for information stored on it for instance according to the requesting entity or the current location.

An electronic license plate is usually based on a passive or active RFID transponder [HC06], which wirelessly transmits the unique vehicle identification number (VIN) and (optionally) further vehicle information to mobile or stationary readers in up to 100 meters distance. However, a thief or criminal as well as the actual driver itself could try to modify or steal an electronic license plate for misinformation or impersonation. Drivers further demand that toll road stations or an arbitrary road user cannot acquire the same amount of information as for instance qualified police forces. Thus, the successful application of electronic license plates requires adequate vehicular IT security measures that thwart unauthorized modification or removal as well as various privacy issues.

4.6.3 Digital Tachograph

The digital tachograph is already a very well considered security-critical component in current vehicles. Digital tachographs record several driving related parameters, which have legal restrictions for public safety reasons such as driving times, rest times, vehicle's speed and vehicle's driven distance. The records are stored in a protected vehicular black box and on the respective driver smartcard for up to 365 days for reviews by public authorities or accident investigations. Digital tachographs are mandatory in the European Union for all vehicles allowed to carry a total weight of over 3.5 tons and vehicles built to carry at least nine passengers, if the vehicle is used for commercial purposes [Eur06].

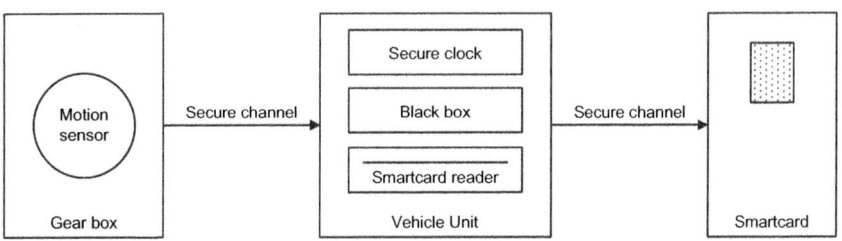

Figure 4.5: General architecture of a digital tachograph system.

As depicted in Figure 4.5, a digital tachograph generally consists of a motion sensor[5] protected inside the gearbox, which transmits its signal to the vehicle unit using a secure channel[6]. The vehicle unit processes the motion signal together with its own clock signal and stores the results internally (black box) and on the driver's personal smartcard. For readout of the internal records or for tachograph maintenance the vehicle unit has to be provided with a properly authorized smartcard. The driver's personal smartcard can also be read using another smartcard reader. However, all smartcard – reader interactions always involve a mutual authentication for providing verification of each others authenticity.

Since tachographs can have serious personal, economic, and legal impacts, they have always been subject to tampering [And98]. The attacker of such a system is usually the vehicle owner (company) or the driver of the vehicle. Hence, an attacker usually has full physical access to the tachograph system, almost unlimited time and almost unlimited trials for a successful attack (cf. Section 5.2.2).

---

[5]The motion sensor provides a signal representative of the vehicle's current speed and/or distance.
[6]A secure channel ensures confidentiality and integrity of the communicated data as well as the authenticity of communication endpoints.

Such hardware attacks can never be prevented at reasonable cost. Again, the goal is to make such an attack infeasible for a rational attacker, that means, the cost of a successful attack shall exceed the potential economic gain (economic security). A digital tachograph can help making tampering much more difficult and thus unprofitable. Therefore, data involved in a digital tachograph system has to be collected, stored and retrieved in a way that unauthorized access and manipulations are virtually infeasible or that tampering attempts are at least detectable. There already exist legal requirements for proper security certification of digital tachographs [Eur02] and manipulations to a tachograph system usually incur a penalty. Clearly, privacy is another crucial security aspect here. There exist four different kinds of smartcards (i.e., driver, forwarding company, workshops, control authority) with four different access authorizations realized by an Europe-wide key management hierarchy.

The security of a digital tachograph system can be provided only by a combination of technical and organizational measures to enable secure channels, to enforce access restrictions, and to detect modifications, manipulations or manipulation attempts. Further reading on current implementions of digital tachographs and discussions on potential weaknesses can be found for instance in [FL06].

4.6.4 Event Data Recorder

The event data recorder (EDR) is a tamper-proof recording device which continuously records several driving related parameters for later accident investigations similar to a black box used on airplanes and trains. In contrast to a digital tachograph, an EDR records additionally to the vehicle's current speed for instance also its current moving direction, its longitudinal and transverse acceleration, vehicle's lightning and safety belt status, and several vehicle control parameters such as current positions of the brake or accelerator pedal [US-06]. Otherwise, an EDR usually maintains only the records of the last few seconds and discards all previous records until an accident stops it, or it starts recording only, when it has registered a significant event that is related to an accident. The records can be read out after an accident using a dedicated reader device or even wirelessly [OnS08].

Even though EDRs are currently mandatory only for public authority vehicles, already millions of private vehicle owners voluntarily installed an EDR to receive for instance special insurance discounts [Off06]. EDRs further provide manufacturers with valuable information about vehicle's performance in a crash event that helps improving the safety mechanisms of future vehicle generations. Since their records can be used in a court, their application introduces first regulations [IEE05, US-06] to standardize the data collected and recorded by an EDR.

However, event data recorders are still the subject of public discussions on legal, technical, and privacy issues [Kow01, Off06], but regardless of the final results, IT security measures are mandatory to enforce access restrictions and prevent any manipulations and on data collected, stored and retrieved using an EDR.

4.6.5 Electronic Log Book

Providing evidence for accomplished trips or critical maintenance operations can be very important for legal restraints, commuting accounts, or warranty claims. Having an integrated electronic service check book and/or driver's log would clearly ease bookkeeping and provide reliable information. Hence, an electronic log book application requires appropriate IT security measures for manipulation and privacy protection.

4.6.6 Road Pricing

Road pricing, toll charging, or drive through payment are further important (legal) vehicular applications. Until now, most toll systems for example either charge a (regionally valid) overall fee whose payment is visually proven by a vignette, or they employ a multiplicity of individual toll stations for usage-based toll charging. However, road pricing based on toll stations is very costly, interferes with the traffic flow, and requires a lot of additional space, whereas overall fees are quite inflexible, require a regularly and non-predictable controlling entity, and may be considered as unfair by charging every road user completely independent of its actual road usage.

Automatic toll charging systems based for instance on V2I communication mechanisms (cf. Section 4.7.3) provide an flexible, cost-effective, easy-to-use, and scalable way for road pricing. First realizations based GPS-enabled on-board units (OBU) and infrared communication with stationary control bridges are already in use [Tol08]. However, next to several functional requirements (e.g., high speed compatibility, multi lane support), automatic road pricing systems have to fulfill also at least the same security objectives as hitherto existing manually operated toll billing systems. This regards for instance several requirements on privacy protection (up to complete driver anonymity), dependable vehicle classification, and particularly, dependable billing to prevent various types of misuse such user profiling, impersonation, billing forgery, or circumvention of billing.

4.7 Vehicular Communication

Until now, vehicular communication (VC) meant mainly in-vehicle communication between different electronic control units (ECU) and their respective sensors distributed all over the vehicle. Communication to the outside world was normally restricted to some vehicle-to-device communication (V2D) interfaces. This usually means proprietary interfaces for workshop fault diagnosis, ECU software updates as well as the weak integration of some of the driver's mobile electronic devices such as cellular phones to enable for instance hands-free calling. However, vehicle manufacturers have already started to integrate first external communication channels that enable their vehicles to wirelessly communicate with their surrounding infrastructure (V2I) using available base stations. Based on integrated GSM[7] receivers, V2I communication is currently already used in some toll billing systems [Tol08]. By also integrating the vehicle's current position, V2I communication enables various so-called location based services (LBS). Future vehicles will even communicate directly with each other using dedicated short range communications (e.g., by making use of the wave band around 5.9 GHz as proposed by the IEEE 802.11p standard [IEE07]), that means communications from vehicle to vehicle (V2V), to exchange for instance information about current emergency events, road conditions, or even for mutual regulation of the right of way.

In the following, for each of the mentioned types of vehicular communication (i.e., in-vehicle, V2D, V2I, V2V) some current and some future applications are introduced, which could be vulnerable to malicious encroachments and hence require appropriate IT security measures for their dependable operation.

4.7.1 In-Vehicle Communication

As shown in Figure 4.6, vehicles of today already contain a multiplicity of controllers and sensors that are increasingly networked together by various internal communication systems with very different properties. In contrast to point-to-point connections, an in-vehicle communication network logically connects several peripherals using the same set of wires. This enables various synergy effects, such that different controllers can share the signal of a single sensor. Networked in-vehicle communication further saves costs and weight, and hence inherently reduces fuel consumption and maintenance requirements. Finally, in-vehicle networks are easy to implement and easy to extend, and are often just mandatory

---

[7]The Global System for Mobile communications (GSM) is the most common standard for digital mobile phones.

to realize several complex networked vehicular applications such as the adaptive cruise control (ACC) or the electronic stability program (ESP). Therefore, in-vehicle communication networks also connect several particular safety-critical components of a vehicle such as breaks, airbags, the engine control, and several steering actuators. Future X-by-wire systems completely rely on in-vehicle networks for correct transmission of their corresponding control commands.

	Sub-bus	Event-triggered	Time-triggered	Multimedia
Exemplar	LIN	CAN	FlexRay	MOST
Exposure	Medium	Great	Acute	Medium
Possible	Loss of	Risk of accident,	Risk of accident,	Data theft,
Harms	functionality	Loss of assists	Loss of control	Loss of comfort

Table 4.1: Endangerment of selected internal vehicular bus systems.

Although internal vehicle communication networks assure safety against several technical interferences [CTG03, HT98, Pol95], they are mostly unprotected against any malicious encroachments. Moreover, as shown in Figure 4.6, the increasing coupling of internal vehicular communication networks with several user-access multimedia busses or even with external wireless networks can cause various unpredictable security risks [Bri07, MZ05], which have hardly been analyzed [PWW04b]. As shown in Table 4.1, the consequences of successful attacks may range from minor comfort restrictions up to acute hazards for the respective occupants and other road users if for instance a systematic malfunction affects an internal real-time communication network, which handles critical steering parameters. Nonetheless, also just a simple malicious door locking could lead to serious consequences for the occupants involved [Ban03].

4.7.2 Vehicle-to-Device Communication

Vehicle-to-device (V2D) communication enables the vehicle to communicate with external devices that actually do not belong to the vehicle itself such as diagnosis or programming devices or various passengers devices such as mobile phones, handheld computers, or mobile audio players. For these, the vehicle normally offers several hardware interfaces (i.e., connecter plugs) and sometimes even wireless interfaces such as Bluetooth [Did03] or W-LAN [IEE07].

V2D communication can be used to connect external test devices for vehicle fault diagnosis, to scan the internal communication, or to contact a particular con-

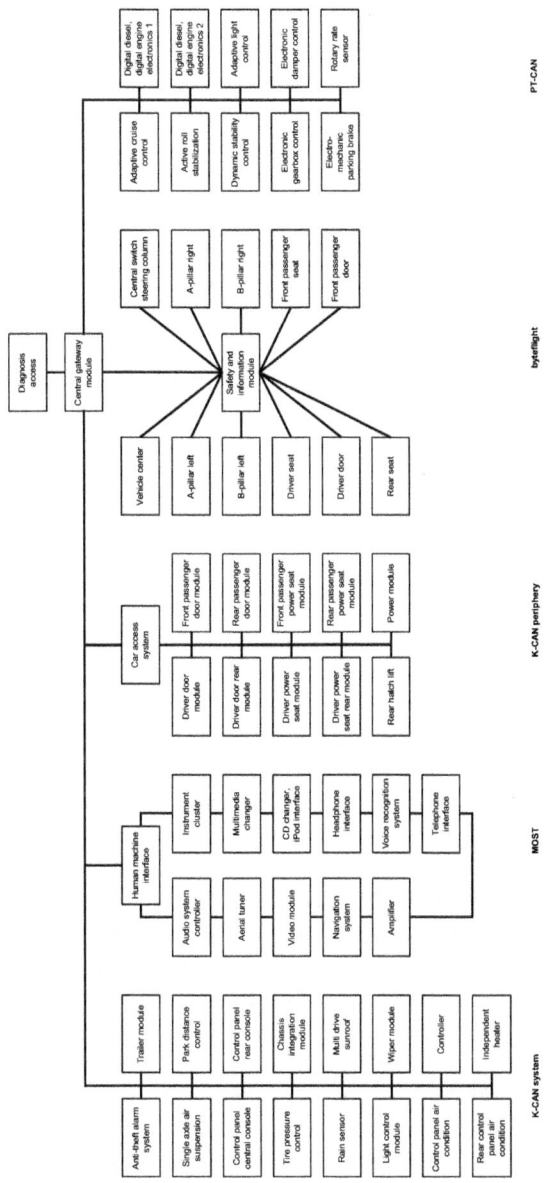

Figure 4.6: In-vehicle communication network of a BMW E65 (7 series) with five different bus systems connecting over 60 ECUs interconnected using a central gateway [HSB+02].

troller. Particularly, V2D communication is used to execute ECU software updates/upgrades (cf. Section 4.4), after-sale feature activation (cf. Section 4.5), or to retrieve afore collected telematics data (e.g., from a digital tachographs or from an EDR). Here, IT security measures are mandatory for the authentication of external devices to verify their access authorization on at least very safety-critical and privacy-critical information. IT security is further required to provide integrity and confidentiality of communicated data, which is of particular importance for instance in case of ECU software updates.

Wireless V2D communication in turn is mainly used to weakly link the driver's mobile electronic devices such as cellulars or mobile audio players with the corresponding vehicle equipment. Therefore, V2D communication enables for instance hands-free calling or automatic muting of the vehicle's stereo system for the cellular, or enables the integration of mobile navigation systems and mobile audio players into the vehicle's head unit and stereo system. Here, IT security measures are mainly required to enforce effective access restrictions to prevent that external (multimedia) devices can affect any internal vehicular system.

4.7.3 Vehicle-to-Infrastructure Communication

Communication between vehicles and infrastructures (V2I), also called car-to-infrastructure (C2I), or vehicle-to-roadside (V2R) communication, means wireless vehicular communication using fixed base stations in the actual transmission area of the vehicle (if available). Today, this means using an integrated GSM or UMTS receiver, which provides a communication link over the public mobile phone network. Some vehicles, however, use an integrated on-board unit (OBU), which is able to communicate with compatible road side units (RSU) using an individual wireless communication technology (e.g., infrared communication or short-range radio communication according to the IEEE 802.11p standard [IEE07]).

V2I communication is currently used to provide drivers with latest traffic information or weather reports that can be integrated into the routing of the actual navigation system, to exclusively receive various OEM premium content, or to enable in-vehicle Internet access. V2I communication is furthermore already part of some automatic toll billing systems [Tol08] used in infrared communication with stationary control bridges, which simplify the payment process and supersede costly conventional toll stations.

Another promising application for V2I communication are electronic traffic signs. A traffic sign recognition (TSR) system warns the driver via head-up display or using the dashboard display for instance when he is driving too fast, he enters a critical area (e.g., road works), or just in case the driver has overseen an important road sign. Current TSR approaches [EAM03, Sie07] depend on the images of a built-in digital camera and verify their results with the corresponding navigation data. V2I-based TSR approaches, however, receive their inputs from electronic road signs that (in addition) wirelessly transmit their (variable) information about actual speed limits, road works, traffic jams, or road conditions to the vehicles passing by. Further traffic sign assist systems will use such TSR-derived information even for active driving interventions such as adjusting the adaptive cruise control (ACC), intelligent shifting, or emergency braking. A vehicle in turn could communicate with an intelligent traffic light system to request for instance setting a certain traffic light, which is otherwise set only on request to avoid unnecessary traffic interruptions. However, the respective receiver (i.e., the vehicle and the vehicular infrastructure) has to detect faulty or bogus information and therefore requires appropriate IT security measures to dependably verify incoming information for validity.

Finally, integrating the vehicle's current location provided by a built-in GPS or GSM receiver (via triangulation) enables various safety, traffic management, business, and infotainment applications widely known as *location based services* (LBS). To prevent potential misuse and manipulations as well as to securely accomplish necessary verifications of data origin, most location based services rely on proper IT security measures to enable for instance secure communication and secure positioning. Since LBS handle many highly personal data, they particularly have to adequately address the passenger's privacy requirements. In the following some exemplary security-critical LBS applications are presented.

Automatic emergency call (eCall). The first popular location-based service is probably the automatic emergency call, mandatory for all new vehicles within the European Union from 2009. As proposed by the eCall Driving Group [Eur08], in case of emergency the eCall system establishes a voice connection directly to a call center initiated either manually by a vehicle passenger or automatically after in-vehicle sensors that have identified an emergency event (e.g., on airbag deployment or via impact sensor signals). At the same time, actual location, available incident or sensitive medical data will be sent to the nearest emergency agency.

Location-based information. Location-based information services might for example allow the driver to find the nearest business of a certain type, for instance, the next fueling station, the next ATM, or available accommodations and restaurants in the immediate vicinity. Optionally, the driver could allow certain location-based incoming information such as traffic reports, obstacle warnings, local points of interest, or some localized advertisements.

Location-based billing. Being able to securely receive the vehicle's current position enables several automatic billing applications, for instance, for toll roads, parking, or congestion charging. Then, drivers could securely pay by an easy acknowledgment within their vehicle (e.g., by sending an 'accept' message) while the operating company or authority is not required to maintain an extra, costly billing infrastructure. More advanced location-based billing applications are car rentals with restricted areas of operation or upcoming pay-as-you-drive insurances [Nor08].

Stolen vehicle localization. As already available with the OnStar [OnS08] technology, LBS allows localization of a stolen vehicle to help police forces pursuing the thief.

Wrong way warning. Several accidents are caused by vehicles driving against the traffic on motorways. Corresponding car radio warnings based on the notices of other drivers are usually already out-dated or at least refer to wrong locations at the time they are broadcasted. A local navigation system, however, could detect the respective wrong lane driving based on the current driving directions in combination with corresponding map data, to warn the against the traffic driving motorist itself and to immediately warn all other road users by transmitting a localized warning message to local traffic information infrastructures.

Fleet management. Modern fleet management systems enable, in addition to vehicle tracking, advanced functionality such as centrally managed routing and efficient dispatching, driver authentication as well as remote diagnosis while gathering details on current driver's status, mileage, fuel consumption or container status.

Floating cellular data. Providing the vehicle's current position, speed, and travel direction to a public traffic management agency enables the realization of intelligent transportation systems (ITS), where every properly equipped vehicle acts as an individual traffic sensor. An ITS then is able to generate for instance detailed traffic statistics, live traffic reports used for general traffic information, parking space availability information systems, or even active traffic interventions by means of adapted speed limits or traffic light controls. In contrast to current ITS approaches that employ digital cameras or embedded road sensors, floating cellular data based ITS do not require extra, extensive and costly, road installations.

4.7.4 Vehicle-to-Vehicle Communication

Instead of exchanging information indirectly using available road side units, vehicles could also communicate directly with each other, using dedicated short range

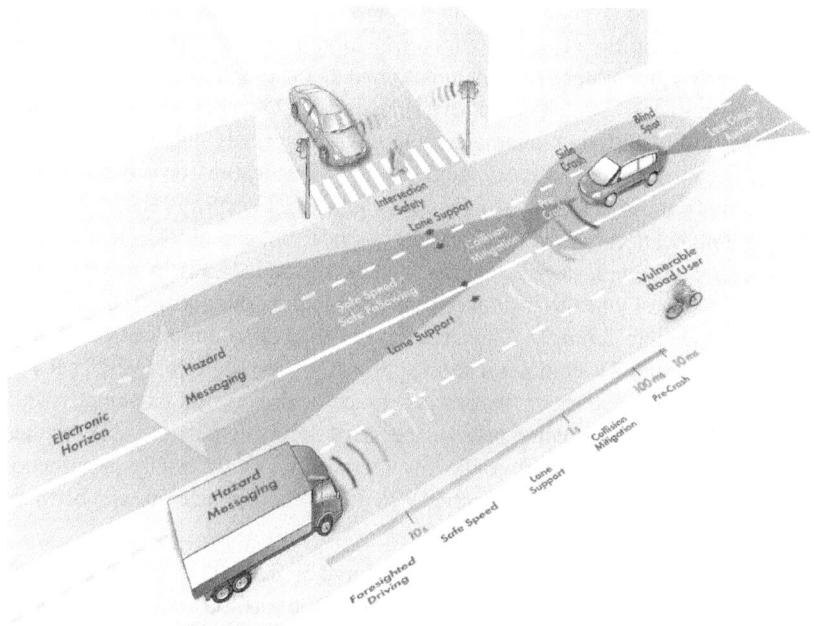

Figure 4.7: Various exemplary V2I and V2V applications pursued by the research project PReVENT [PRe04] such as emergency event warning, collision avoidance, lane (change) assistance, and right of way regulation.

communication (DSRC)[8] technologies as currently implemented for instance by the IEEE 802.11p [IEE07] standard. This kind of communication is called vehicle-to-vehicle (V2V) or car-to-car (C2C) communication. In contrast to V2I communication, a typical V2V communication scenario has only a very individual and situational (ad-hoc) set of senders and receivers. Even though V2V communication is currently rather a vehicular technology for the future and is still a recent field of research [Car05, CVI04, Net04, PRe04, RPH06, SAF06, SeV06], as shown in Figure 4.7, it could enable a lot of new safety-enhancing applications, probably more than any other vehicular technology has done before. In fact, having V2V communication once sufficiently deployed in the field would be a technical revolution in the automotive development.

---

[8]DSRC is a wireless communication technology specifically designed for V2I and V2V applications using a wave band around 5.9 GHz in the US [IEE06a] and 5.8 GHz in Europe and Japan [EN 04].

However, since V2V communication works even without a third instance that could act as an additional control and verification instance, V2V communication is particularly vulnerable to data interception, data manipulation, masquerading or fake messages [ABD+06]. Particularly, since most V2V applications often handle very safety-critical driving parameters, effective IT security measures, which enable for instance mutual authentication, integrity verification, or non-repudiation are absolutely indispensable. Another very important security related aspect of V2V communication is privacy, particularly since the amount of personal information being collected, stored, analyzed and aggregated already by non-vehicular applications (e.g., by health insurances, communication providers, or financial institutions) rises threateningly. Thus, emerging vehicular communication and particularly V2I and V2V communication applications require effective IT security measures that fairly consider the privacy issues of drivers and passengers while enforcing the least information policy[9]. Some of the many possible V2V applications that could be affected by malicious encroachments are briefly described in the following. However, several further V2V (and V2I) applications can be found in [DOT04].

Emergency event warning. Vehicles that observe an anomalous event such as an accident, a motorist driving on the wrong side of the road (i.e., a so called "ghost driver"), or potentially dangerous road conditions (e.g., heavy rain or ice) could immediately broadcast a warning message to all vehicles in their surrounding area to inform all drivers in the corresponding area in advance. Adaptive cruise control (ACC) systems could furthermore use such warning messages to automatically adapt the vehicle's current speed.

Lane change/merge assistance. Before changing the lane or lane merging, V2V-enabled vehicles could warn each other in case they are currently driving just side by side, or could even automatically negotiate who goes in front and who stays behind.

Right of way regulation and cooperative driving. Similar to the lane change assistance, V2V-enabled vehicles could also warn each other in case a vehicle is going to (accidentally) take another vehicle's right of way, or could automatically negotiate the right of way in case a traffic situation requires mutual coordination (e.g., roads with equal rights).

Collision avoidance systems. In co-operation with a right of way regulation mechanism and a lane change assistance, a collision avoidance system would continu-

---

[9]The least information policy requires that components that are not under full control of the actual user shall be able to collect, store, and reveal the user's private information only to the extent as absolutely necessary for their correct operation.

ously monitor the current environment of a vehicle to warn the driver if another vehicle could cause an collision. It further could automatically negotiate a mutual avoidance maneuver to prevent an impending accident.

Priority signal. Emergency vehicles or police cars could broadcast a priority message to automatically set traffic lights and adapt right of way regulations in order to reach their emergency destination safely and quickly.

Cooperative Adaptive Cruise Control. Coupling a group of vehicles together for automatically coordinated maneuvering (*platooning*) could make driving more comfortable and safe and could enable even driverless vehicle operation.

Multi-hop data routing. Using vehicles as nodes in a dynamic ad-hoc or mesh network would enable various multi-hop routing protocols to enable vehicles to exchange information over their actual transmission range, but without relying on available infrastructures.

4.8 Protection of Safety-Critical Applications

Advancements in vehicular safety introduced various active and passive safety measures and considerably increased the reliability of vehicles and vehicular systems. Nevertheless, automobiles still are one of the most dangerous application day-to-day used by billions of people [PSS⁺04]. And, as already described in Section 4.7, today's highly complex vehicular safety systems such as anti-lock braking (ABS) or electronic stability programs (ESP) already had to abandon the previous automotive engineering approach of strictly separating safety-critical systems from other rather critical vehicular systems. Thus, an ESP system for instance, is already manifoldly interconnected with various sensors, actuators, and microcontrollers using several different in-vehicle networks. Future, V2I- and V2V-based safety applications moreover rely on various open external interfaces, which makes them even more susceptible to malicious encroachments.

Hence, it is not even necessary to use the terrorist attacker that encroaches into another's antiskid system as dramatic example to alert the need for protection of safety-critical vehicular applications. In fact, almost *all* vehicular IT systems could quickly become safety-critical and hence susceptible to security issues. Even vehicular applications that seem, at least at the first glance, to be fully non-critical could have serious impacts on driving safety. It may, for instance, suffice to imagine some simple (malicious) manipulations that cause a suddenly uncontrolled spattering windshield washer (without the wipers going on), a full interior lighting at night (while turning the headlights off), or the car radio going abruptly on full blast—all during a critical traffic situation. Hence, even a simple malfunction of

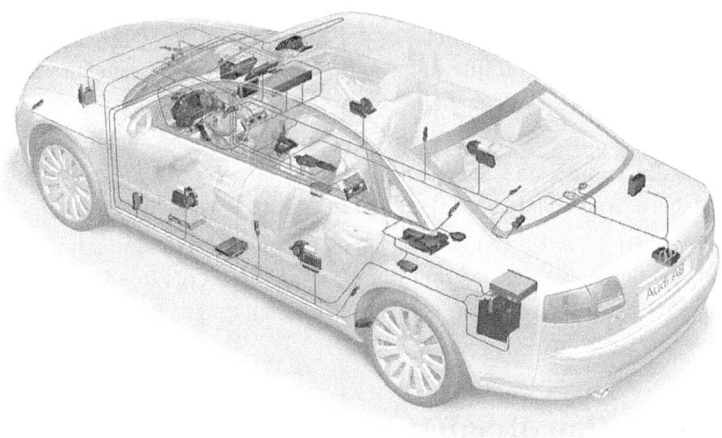

Figure 4.8: In premium vehicles, one can find up to 80 processors that are interconnected by up to five different bus types and up to several 100 megabytes of embedded code providing more than 2000 individual functions [Bro06]. (Image: Audi AG [Aud08])

an usual electronic door look system can endanger lives, if it hinders the occupants leaving the car in a critical situation [Ban03]. Alone, the continuously increasing interconnection and interdependency of most vehicular IT applications virtually makes all of them also subject to IT security issues. Especially, since attacks that simply target the availability of certain vehicular controllers and functionalities often can be mounted very trivially. This is due to the fact that most vehicular applications are indeed developed to face random technical failures (e.g., by verifying checksums or ensuring high redundancy), but almost never consider a malicious human attacker that uses a certain functionality in a syntactically correct way, but in a bad faith.

Hence, according to statements of the avionic safety engineer Driscoll [Dri02], safety-critical vehicular IT systems actually demand at least the same, if not even higher, security and reliability standards as currently usual in aviation. Since, for example in case of a safety-critical failure in avionics, jet pilots usually have both several independent possibilities to steer the affected airplane (e.g., engines, aileron, rudder, or elevator) and considerably more time to react effectively. Failures of a safety-critical vehicular IT system (e.g., due to malicious manipulations), in contrast, can often quickly cause serious consequences since an ordinary car driver usually simply does have neither enough time nor sufficient options to react effectively.

4.9 Privacy Protection

Even though, various privacy implications of several security-critical applications were already mentioned in the previous sections, privacy protection is such a major concern in many vehicular IT systems, that it will be again, separately and concretely, considered in this section.

Privacy protection provides that the usage of vehicular IT systems do not violate privacy policies and thus prevents harm, embarrassment, inconvenience, or unfairness to any party whose data is processed [Shi00]. Moreover, in compliance with the relevant applicable laws, each vehicle user should be able to determine for himself whether or to which extent it shares personal information with others. This concerns in particular the least information policy [GLLD05] such that components not under full control of the user shall be able to collect, store, and reveal user's private information only to the extent essentially required for its operation.

Today many everyday life activities are already used to gather various personal information. Public cameras monitor public places, shopping cards gather information regarding the individual purchasing behavior, while Internet and telecommunication service providers collect various information regarding the individual communication behavior (e.g., time, destination, extent). Ongoing advances in information technology (e.g., computing performance, storage and transmission capacities) concurrently enable to store, share, aggregate, and analyze huge amounts of personal information gathered in this way throughout the world. This therefore already suffices to determine for instance someone's age, purchasing behavior and purchasing power, education, employment, health, leisure activities, location, or religion that means personal information that most people probably would not like to be publicly available. Thus, it would be desirable to prevent any further privacy impairments by the deployment of new vehicular IT applications, apart from the fact that most vehicle user probably will object vehicular applications, which requires them to disclose significantly more private information as hitherto usual. However, driving a vehicle has never been fully anonymous. For instance, most vehicles have always attached a publicly visible unique licence plate. Thus, vehicular applications can built up on existing privacy agreements current drivers already accept, but must not further endanger the privacy of the involved individuals. This, however, requires an awareness for potential privacy implications already during the design of a vehicular application and before. That means a sensitive consideration for all information created, received, processed, or provided by a vehicular application with regard to their actual benefits against possible costs on privacy. Once, the processing of a sensitive information is inevitable, the following questions demand adequate answers.

- Who is the owner of certain vehicular information?

- Who is allowed to access which information in which context?

- Who is what allowed to do with the information?

- Exists a possibility for a general, subsequent, or temporary revocation?

Particularly, for possible legal procedures or any further usage, it is important to know, who is the rightful owner of a certain vehicular information (e.g., the OEM, the driver, the vehicle owner, the service provider) before defining any access and usage rights. In the following, several possible sensitive vehicular information are referred, which require adequate privacy protection.

In-vehicle personal information such as address or phone book entries, personal data (e.g., documents, e-mails), or recent navigation destinations processed by vehicular infotainment applications as well as *emergency information* such as current vehicle occupants their personal physiological and medical data transmitted during an automatic emergency call in case of an accident.

Vehicle driving state parameters such as current speed, moving direction, longitudinal and transverse accelerations, *vehicle driving control parameters* such as positions of the brake or accelerator pedal, or the current steering angle, and *vehicle safety state information* such as vehicle's lightning or the safety belt status processed by potential event recorders, digital tachographs, or for V2I, V2C, and internal safety applications.

Vehicle diagnosis information such as vehicle's technical configuration, ECU errors and warnings, maintenance information, current milage processed by (remote) diagnosis, log book or legal applications.

Vehicle identity information such as vehicle's pseudonym(s), license plate, driver licence, current occupants, and basic technical data, *vehicle location information* such as vehicle's absolute (i.e.,) or relative position (e.g., right lane), and *vehicle environmental sensor information* such as weather or road conditions, parking space allocation, identified traffic risk (e.g., accidents, work zones, emergency brakes) processed by various traffic management, road pricing, and VIC applications according to the corresponding access and usage rights.

For possible privacy-preserving mechanisms, protocols, and anonymization services (cf. Section 8.6.3), anonymity has to be quantified in order (i) to assess and to compare different anonymization approaches, (ii) to evaluate the effectiveness of different privacy attacks, and (iii) to be able to quantify potential losses and gains of anonymity. Fortunately, there exist some entropy-based anonymity metrics that can be applied to anonymization approaches and corresponding scenarios [Día05]. Accordingly, a discrete random variable X represents a certain pseudonym that can be correctly linked to a certain subject i (e.g., to an certain individual or to a certain vehicle) from a set of N possible subjects that are involved with the probability $p_i = Pr(X = i)$. Then, the current entropy $H(X)$ of the corresponding pseudonym can be calculated as follows.

$$H(X) = -\sum_{i=1}^{N} p_i \log_2(p_i)$$

Thus, $H(X)$ represents the distribution of probabilities that link a set of pseudonyms correctly to a set subjects obtained by an adversary after he has deployed an attack on the given anonymization system. The highest entropy that is possible for a given set of N possible subjects corresponds to the entropy of a uniform distribution so that all involved subjects are indistinguishable for an adversary. Then H_{max} denotes as follows.

$$H_{max} = \log_2(N)$$

Now, the amount of pseudonym-subject-correlation information revealed by an adversary that is represented by $H(X)$ can be compared with H_{max}, the maximum entropy possible. Thus, the normalized *degree of anonymity* $d = [0\ldots1]$ that is provided by a given anonymization system can be defined as follows.

$$d = 1 - \frac{H_{max} - H(X)}{H_{max}} = \frac{H(X)}{H_{max}}$$

Hence, the effective degree of anonymity d can be increased by two factors, namely by increasing the number of involved entities N or by increasing the uniformity of the probabilities p_i. That means, ideally all subjects are indistinguishable for an adversary, so that $p_i = 1/N$ for a given set of N involved entities.

5 Attackers and Attacks in the Automotive Domain

This chapter identifies several current and future attacking scenarios and classifies corresponding attackers characteristic in the automotive domain. It further presents some feasible logical and physical attacks that could be conducted by the afore classified attackers. Parts of this chapter are based on published research in [BEWW07, PWW04b, WWW07].

5.1 Attackers in the Automotive Domain

As shown in the previous chapter, today's attackers in the automotive domain can have various intentions to attack a particular vehicular IT system. They could try to gain individual benefits, to somehow exploit an attack commercially, or they just handle with malice aforethought. Hence, according to Chapter 4, which introduced several security-critical vehicular IT systems, some of the most common reasons for attacking a today vehicle might be to:

- steal a vehicle as a whole,

- steal a certain valuable component (e.g., the navigation system),

- modify a certain component in order:
  - □ circumvent effective hardware restrictions or limitations (e.g., maximum speed unlocking),
  - □ circumvent effective software restrictions or limitations (e.g., chip tuning),
  - □ activate after-sale features without authorization,
  - □ manipulate driving records (e.g., tachograph, milage counter),

- violate other's privacy or to illegally use other's resources (e.g., phone hack),

- steal a manufacturer's expertise or intellectual property (e.g., to sell counterfeits).

With the introduction of further advanced and complex, future vehicular applications such as electronic license plates, event data recorders, vehicular communication, and copyrighted infotainment, potential misuse and attack incentives will rather increase further than decrease. Hence, again in reference to Chapter 4, future attackers furthermore could try to:

- impersonate (e.g., electronic licence plate, electronic identification),

- circumvent effective digital rights management systems,

- manipulate vehicular communication in order to:

    - gerrymander (e.g., misuse of priority signal),

    - disturb communication (e.g., applying jamming transmitters or repeated bogus message injections),

    - harm people (e.g., trying to cause accidents by malicious encroachments on right of way regulation mechanisms),

    - to wiretap private information (e.g., unauthorized tachograph readout).

According to their individual access perimeters, their technical, financial, and knowledge resources, attackers in the automotive domain can be classified into at least four different groups. As depicted in in Table 5.1, attackers of vehicular IT systems are either internal attackers of different power (I_x) or external attackers (E_0). An *external attacker*, such as a thief or an attacker on wireless vehicular communications, tries to attack a vehicular IT system from the outside without being an authorized or legitimate user of the respective system (*outsider*). He may have considerable technical expertise and some appropriate tools, but normally has only very limited or no physical access to the attack target. Further, he normally has only limited time to successfully mount an attack. Hence, an external attacker generally can execute only attacks which exploit logical weaknesses (cf. Section 5.2.1) and which are practicable virtually in realtime (*online attack*). Due to these limitations, reliable protection against external attackers on vehicular IT systems is usually feasible.

In contrast to an external attacker, an *internal attacker* normally is an authorized or legitimate user of the respective vehicular IT system he attacks (*insider*). Hence, internal attackers usually try to access or use a particular vehicular IT system in a way which exceeds their actually granted authorization. In order to successfully mount an attack, internal attackers of an vehicular IT system moreover have generally almost unlimited time and full physical access to the attack target. Hence,

	Attacker I_1 Internal Class I	Attacker I_2 Internal Class II	Attacker I_3 Internal Class III	Attacker E_0 External Class 0
Exemplary attackers	Driver, owner	Motor mechanics, backyard garage	Organized crime, rival, academia	Thief, V2I or V2V mischief
Physical access	Limited to resp. skills	Extensive, but not unlimited	Virtually unlimited	None or only very limited
Technical resources	Generally low	Medium to high	Very high	Varies, usually low to medium
Knowledge resources	Generally low	Medium to high	Very high	Varies, but can be high
Financial resources	Low	Medium	Very high	Generally low
Reliable protection	Mostly feasible	Varies, but still feasible	Only by econ. security	Mostly feasible

Table 5.1: Potential attackers in the automotive domain classified according to their access perimeters, technical, financial, and knowledge resources.

they can try to mount almost any feasible attack, which exploits logical (cf. Section 5.2.1) or physical weaknesses (cf. Section 5.2.2) without having to fear to become detected, backtracked, or locked out (*offline attack*). However, internal attackers differ in their available technical, financial, and knowledge resources and thus can be classified into at least three different groups from I_1 to I_3.

The group of I_1 attackers refers to individuals such as the car owner or the actual driver, which try to breach the security of a certain vehicular IT system. For that, they normally have only limited technical, financial, and knowledge resources. Hence, their actual physical access to the attack target is also practically limited according to their usually limited manual skills, expertise, and technical equipment. Because of the in all respects restricted attack potential, reliable protection of vehicular IT systems against I_1 attackers should normally be feasible.

The group of I_2 attackers refers to skilled (OEM) garage employees and mechanics, which usually act on behalf of the respective vehicle owner or driver to breach the security of a certain vehicular IT system. They generally have appropriate tools and necessary technical expertise and are furthermore mostly endued with a certain amount of insider information. They would even invest up to a medium amount[1] of money, if an attack promises appropriate revenues, or if a once successfully accomplished attack can easily be applied to other vehicles afterwards. Because of the generally quite large attack potential, reliable protection of vehicular IT systems against I_2 attackers requires considerable efforts and would be feasible only by costly (hardware-based) protection measures.

The group of I_3 attackers finally refers to concurrent manufacturers, counterfeiters or the organized crime that may have immense technical, financial, and knowledge resources limited only by the potential economic gain. Since revenues on the counterfeit market or a certain competitive advantage can be worth several millions of US$, an I_3 class attacker can draw from latest research knowledge, is able to access expensive high technology equipment and to employ several people to successfully mount an attack. Thus, an I_3 attacker is potentially able to break any IT security measure, if it would provide a sufficient economic gain. In fact, the success of an I_3 attacker can be restricted only by economic security such that the total cost of a successful attack exceeds its potential economic gain. Hence, a single successful attack on an automotive device must not scale to break also all other devices, for example, by revealing a global identical secret. However, effective IP protection for instance cannot be realized solely by built-in vehicular IT security measures and hence requires additional measures, which for instance would provide an evidence if a company reuses competitor's intellectual property. Nonetheless, even well equipped able academical researchers may try to breach the security of vehicular components and mechanisms to reveal potential security risks and to enhance their scientific reputation.

5.2 Attacks in the Automotive Domain

If an attacker in the automotive domain (cf. Section 5.1) attacks a vehicular IT system, that means, if he tries to breach the security of a particular vehicular IT system, he usually tries to exploit a certain vulnerability or just attacks the weakest link of the effective security protection measure. Similar to attacks on common IT systems, attacks in the automotive domain can be accomplished during the development, the production, and during the actual deployment of a vehicular IT system. However, due to several typical characteristics in the automotive domain (cf. Section 6.4), attacks on vehicular IT systems often differ from attacks on general-purpose computer systems. This applies particularly to attacks on vehicular IT systems when they are finally in field use within the vehicle. In contrast to

---

[1]A "medium amount of money" means financial resources available in a typical garage, that are generally less than 100,000 US$.

most attackers of common computer systems, attackers of automotive IT systems for instance usually have full *physical access* to the target device. Hence, without proper protection measures, an attacker is able to manipulate or even to replace almost every built-in component and can almost arbitrarily manipulate its actual environment and (physical) inputs.

Attacks on vehicular IT systems are further normally *offline attacks*, where an attacker has virtually *unlimited time* and virtually *unlimited trials* to successfully mount an attack. Hence, the attacker can undisturbedly mount almost any feasible attack without having to fear to become detected, backtracked, or locked out. Due to the typical long vehicular product life cycle and the slow update rate, he will further always find a vulnerable target, even if a certain security issue is already known and fixed for a long time.

According to the access perimeter of the respective attacker (cf. Figure 5.1), vehicular attacks in general can be classified in logical and physical attacks. Logical attacks can be accomplished by an external attacker E_0 and by all internal attackers I_x, while only being able to access available logical interfaces as well as external communication. However, an E_0 attacker certainly has access to much less logical interfaces than an internal attacker I_x and thus much less attack potential. Physical attacks, in contrast, can be accomplished only by an internal attacker I_x since they require direct physical access to the respective attack target or its actual physical environment. However, the depth and complexity of feasible physical attacks differ by orders of magnitude between for instance an I_1 and an I_3 attacker.

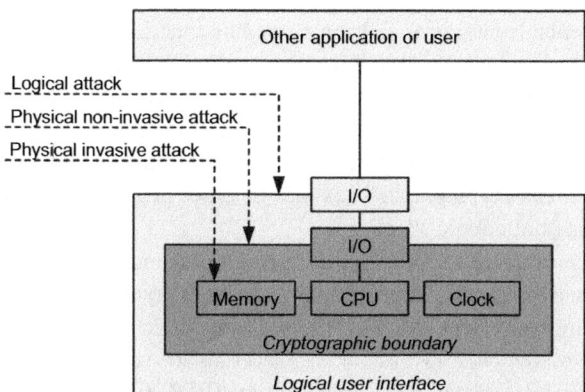

Figure 5.1: Different maximum attack perimeters of a logical attack, a physical non-invasive attack, and a physical invasive attack on a security-critical vehicular application.

However, since most concrete attacks are heavily dependent on several very application-specific and implementation-specific details, this chapter presents in the following only a basic overview of possible attacks feasible in the automotive domain without giving too many details.

5.2.1 Logical Attacks

Logical attacks are attacks via available logical interfaces and external communication, include passive (e.g., wiretapping) as well as active (e.g., privilege escalation) encroachments. Here, an attacker could for instance try to exploit available external communication interfaces such as used by door locking or VC applications. An internal attacker could further try to exploit internal logical interfaces for instance provided by the vehicular user interface for software updates or feature activations.

The following exemplary attack procedures are quite similar to (application) attacks in general-purpose computing, since they assume a quite similar attacker model.

Cryptographic Attacks

Cryptographic attacks or cryptanalysis[2] tries to break the security of IT systems while exploiting potential conceptual weaknesses or while just "brute forcing" the underlying cryptography of cryptographic algorithms or (cryptographic) protocols.Therefore, an attacker can analyze intercepted communication (passive) and utilize all available interfaces and corresponding functionality as actually provided (active). Whereas (usually recognized as computationally infeasible) brute force attacks try to break a cryptographic system by systematically trying all possibilities (e.g., an exhaustive search other all possible keys in order to decrypt a message), cryptanalysis tries to exploit specific theoretical or implementational weaknesses or flaws to break cryptographic schemes in order of magnitudes faster than the corresponding brute force attack.

Since at least current standardized cryptographic algorithms and protocols are developed and verified by internationally accepted cryptographers, attacks on the actual cryptography very seldom represent the weakest link in order to breach the security of vehicular IT system. However, as the recent research publications on the automotive-related KeeLoq algorithm [BDI+07] or the wireless encryption standard WEP [TWP07] exemplify, the usage of proprietary, little verified,

---

[2]However, cryptanalysis is not restricted to logical attacks only.

or outdated algorithms and protocols can nevertheless result in serious security vulnerabilities.

Software Attacks

Software attacks refer to all attacks based only on the unintended utilization of the provided software interfaces and functionality, but do not include attacks on underlying cryptographic schemes (cf. cryptographic attacks). Hence, the actually provided functionality is often even a superset of the intended functionality (i.e., provided functionality $\supseteq$ intended functionality). Software attacks particularly exploit flaws and vulnerabilities in design or implementation[3] such as:

- integer, heap or buffer overflows,

- race conditions or infinite loops,

- corrupting input and output data,

- dependency corruption,

- resources exhaustion,

- enforcing restarts or resets,

to cause any kind of unintended behavior (e.g., a program termination, or a privilege escalation) that can be used for an attack (e.g., code injections, denial of service). Software attacks, however, are errors in the design or implementation and are either failures in meeting the corresponding requirements or results of insufficient, inconsistent, or incorrect design specifications. However, since today software-driven vehicles already use up to several 100 megabytes of embedded code providing more than 2000 individual functions [Bro06], it becomes more and more difficult and costly to evaluate such amounts of software properly. With regard to the world of general-purpose computers, software attacks based on exploits and vulnerabilities will, without further measures, become one of most critical security issues in the automotive domain. After all, also all the subsequenting communication attacks are in the end even "higher-order" software attacks that means weaknesses in design or implementation.

---

[3]In [HM04], Hoglund et al. give a good general introduction to the subject of software attacks.

Communication Attacks

Attacks on vehicular communication are probably the primary challenge for the most future VC-based applications (cf. Section 4.7). As shown in Table 4.1, the consequences of successful attacks may range from minor comfort restrictions up to acute hazards for respective occupants and other road users.

Communication attacks comprise active and passive attacks on all communication channels accessible by the respective attacker. This includes not only external attackers, but also malicious insiders, which try employ their equipment in a malicious manner or employ even modified equipment to mount an attack. Passive attacks on VC usually eavesdrop communicated information for instance to perform an offline attack, but do not interfere with actual communication procedures. In contrast, active attacks on vehicular communications directly interfere with communications while systematically intercepting, manipulating or injecting communicated messages. In the following, a brief overview of feasible attacks on vehicular communication is given. However, more detailed analyses can be found in several VC-related research articles [ABD$^+$06, BE04, DGL$^+$02, PGH06, RH07, RPH06] and in the publications of VC-related research consortiums [Car05, CVI04, GST05, Net04, SAF06, SeV06] respectively.

Eavesdropping or *wiretapping* means passive attacks that listen to other's communications without the authorization or even the knowledge of the communicating parties. The monitored information then can be analyzed (offline), for example, to disclose sensitive information or can be used to mount further attacks (e.g., replay attacks). If a potential information leakage exists, eavesdropping of freely available communication that means communication that is assumed to be protected or non-critical can be used to retrieve utilizable information or to prepare another attack (e.g., *aggregation attacks* based on a specific communication context, message frequency, or origin of a message).

Manipulation attacks on vehicular communications are all active malicious modifications, suppressions, or substitutions to communicated information or communication flows in order to, for example, gain unauthorized access, to enable further attacks (e.g., masquerading attacks), or to merely interfere or sabotage vehicular communications.

Injection attacks or *insertion/fabrication attacks* introduce false messages into a vehicular communications, which pretend to origin from a valid source. Therefore, an attacker can generate new valid messages or retransmit afore intercepted (modified) messages (cf. eavesdropping).

Replay attacks are man-in-the-middle attacks in which a valid communication (or parts of it) is maliciously repeated or delayed. Therefore, the attacker eavesdrops or intercepts the communicated information to reuse it in a malicious retransmission. Replay attacks can be used for instance to bypass simple authentication mechanisms that solely rely on the transmission of a fixed (i.e., repetitively used) secret. Early vehicular remote door lock systems (cf. Section 4.2.2), for example, which often only verified a static password were hence particularly vulnerable to replay attacks.

Relay attacks or *Mafia fraud attacks* are real-time attacks against identification or authentication schemes, where the attacker acts as man-in-the-middle forwarding all messages verbatim between a valid sender and the valid receiver. However, the sender usually is either not aware of this communication or is made to believe that the attacker is a valid receiver. As briefly introduced in Section 4.2.1, a relay attack can be used for instance to illegally access a vehicle while covertly relaying the authentication information from the distant original vehicle key to the actual vehicle (cf. Figure 4.2). A more sophisticated relay attack is the wormhole attack [HPJ02] where an attacker controls at least two nodes in a network such that one node directly tunnels messages recorded in one location to the second node that resends them into the network in another location.

Impersonation and Masquerading attacks are special manipulation or injection attacks on vehicular communications, where the attacker pretends to be a certain valid communication party in order to gain other's privileges or to act with someone else's identity. These attacks could be used in several payment applications (e.g., toll applications), to abuse other's identities or authorities (e.g., police or emergency priority signals), or to hide the attacker's identity for further attacks.

Privacy violations are a very critical characteristic often inherent in most VC applications and threat the privacy policies of any involved party (usually the driver). Typical privacy threats that could be induced by VC are for instance tracking of vehicle's location, monitoring the driver's behavior or simply wiretapping unprotected communicated sensitive information (cf. Section 4.9).

Disclaiming of liability refers to methods that enable a malicious party involved in a vehicular communication to conceal or deny that he is the origin of certain information. Disclaiming the liability could be imaginable, for instance, to conceal misdemeanors recorded by digital tachographs, to subsequently deny payment authorizations, or to unverifiably inject false or bogus messages in several V2X applications (cf. Section 4.7).

Denial-of-service attacks or jamming attacks finally refer to all attacks that logically considerably restrict (or prevent) the availability of vehicular communications in a malicious manner. In contrast to physical DoS attacks (cf. Section 5.2.2), logical DoS attacks can be performed for instance by inducing fault attacks or by excessively invoking a critical function. Since most DoS attacks normally can be accomplished without compromising specific security or even cryptographic measures, at least local DoS attacks can be comparatively effective.

5.2.2 Physical Attacks

As mentioned in the section's introduction, physical attacks assume that the attacker has direct access to the attack target, that means that the device, which has to be protected, operates in an insecure environment. Hence, the security depends, additionally to its logical capabilities, particularly on its physical construction and corresponding physical protection measures. However, physical attacks are often independent of the corresponding software (countermeasure) implementation by directly attacking the hardware to:

- access internal (hardware protected) secrets (e.g., cryptographic keys),

- disable (physical) security measures (e.g., intrusion sensors or security filters) to enable further attacks,

- introduce physical signals (e.g., anomalous voltages, strong radiation) to to enable further attacks.

Thus, a physical attacker could for instance try to disable the signature verification mechanism for software updates in order to install unauthorized software. Or an attacker could try to discover a (hardware-based) secret decryption key used to protect valuable or personal information such trade secrets (e.g., engine characteristics), intellectual property (e.g., vehicular program code or data), or certain driving records. A physical attacker could also simply try to modify different vehicular functions that (so far) solely rely on software or logical protections (e.g., feature activations or milage modifications).

There exist only a few comparable scenarios, where the security-critical device operates in a hostile environment. So far, only smart-card applications have to thwart similar attacks. In the following, several exemplary physical attacks are examined that can be distinguished in passive and active attacks as well as in invasive and non-invasive attacks. An active attack directly attempts to alter internal resources or affect their operation, whereas a passive attack attempts to make use of information from the system during normal operation, but does not directly affect internal resources. Non-invasive physical attacks can access or manipulate the actual physical environment including the corresponding physical interfaces, but cannot access device internals behind a so called *cryptographic boundary* [FIP02a] which encloses all security-critical components (e.g., processor, memory, random number generator, clock). Invasive attacks in turn, are able to attack even device components behind the cryptographic boundary. Moreover, invasive attacks usually physically harm the attack target in an irreversible manner. Physical attacks furthermore can be accomplished during (normal) operation of the attack target (dynamic attack) or while the attack target is switched off (static attack). Further reading on physical attacks, particularly on monitoring, sidechannel, fault and penetration attacks, can be found in [KK99, SLP06, Wei00].

Monitoring and Sidechannel Attacks

Monitoring and sidechannel attacks mean the passive, non-invasive interception and examination of all channels susceptible to the attacker that allow him to breach the security of the respective device (e.g., to reveal an internal secret key). This does not only mean information externally communicated, but also information leaked by the actual physical implementation such as the respective execution time, power consumption, or electromagnetic emission, which enable for instance timing attacks or attacks based on power analysis or electromagnetic emanation analysis [Sma00]. Due to their non-invasive, non-harming manner, and the rarely available and the yet more rarely implemented proper countermeasures, sidechannel attacks are a particularly dangerous attack method against many vehicular IT security mechanisms.

Denial of Service Attacks

Physical denial of service attacks refer to active, invasive and non-invasive attacks on the physical availability while preventing or considerably delaying access to critical resources, communication, or functionality. This can be realized by physical jamming, malicious deletion, deactivation or malfunction enforcing, or even by physical destruction. For that, an attacker could use for instance powerful electromagnetic fields or heavy noise inductions (non-invasive attacks), or he could try to short-circuit, detach, or shield (invasive attacks) the corresponding component.

Fault Attacks

Fault attacks or perturbation attacks are active, non-invasive attacks that expose the respective attack target to an anomalous operating environment to induce faults that could disrupt or modify the execution of some critical instructions or critical values. Therefore, the attacker exposes the target for instance to extremely high or low temperatures, induces power or timing glitches (if an external clock signal is used), or uses high energy radiation (e.g., lasers, X-rays, or ion beams) to bypass for instance a critical authentication procedure.

Penetration Attacks

Penetration attacks are active, invasive attacks that intercept internal communications, readout internal memories, or monitor any internal processing behavior to discover internal structures, secrets, and functionality. They are executed by sophisticated microprobing techniques or even by depackaging of the respective microchip followed by a layout reconstruction via high-resolution photographs or electron microscopy. Since penetration attacks allow to reveal the technical principles and internal operations of the corresponding component, it is a common technology for *reverse engineering.*

Modification Attacks

In contrast to penetration attacks, modification attacks additionally enclose arbitrary manipulations or even complete substitutions of any internal component (e.g., processor, memory, clock, random number generator) as well as arbitrary manipulation of internal communications.

Exploit of Testing Functionality

Many IT devices include different test, verification, or diagnose functionality, which is not properly deactivated and hence can be misused for an attack. Once such mostly undocumented functionality has been discovered, it can be exploited for the disclosure of internal information or for modifications of the functionality or for modifications of internal values of the corresponding device.

5.2.3 Further Attacks

As done also in general-purpose computing, an attacker in the automotive domain can also try to exploit potential organizational weaknesses or weak policies. *Social engineering*, for instance, refers to attacks and procedures used to gather useful (confidential) information while manipulating or fooling people, or exploiting typical human behavior. Examples for technical social engineering are phishing techniques, Trojan horses, or mostly the vulnerable general IT infrastructure of the respective vehicular manufacturer. Non-technical social engineering examples are bogus workmen, going through trash bins to find valuable information ("dumpster diving"), or industrial espionage in general. Unintentional information disclosures in cooperation with external contractors is another important security vulnerability that can be thwarted only by strong access control policies along with additional legal commitments. Finally, attacks that directly assault the respective driver (e.g., car hijacking or housebreaking) or just pick up a vehicle as a whole are hardly to thwart solely by IT security measures, but have to be considered in a thorough security analysis and a holistic security design.

6 Security Analysis and Characteristical Constraints in the Automotive Domain

Based on the threat analysis in Chapter 5, this chapter describes how to identify the particular security objectives of the entities involved in typical automotive security-critical applications. It then describes how to deduce the corresponding security requirements to fulfill the before identified security objectives. It further indicates characteristical technical and non-technical constraints as well as characteristical advantages that have to be faced while establishing IT security in the automotive domain. Parts of this chapter are based on published research in [PW08, WWW07].

6.1 Security Objectives Analysis

To guarantee road safety and operational reliability of vehicles and to sufficiently protect automotive business models, legacy and comfort applications that are based on the security of the vehicular platform, the following overall security objectives (SO) are reasonable. Note that these security objectives are additional objectives, in addition to other mandatory, but not security related objectives such as usability or scalability (cf. Section 6.4.1 and Section 6.4.2, respectively).

Authenticity. The origin of a vehicular information (e.g., message, program, or data) or a vehicular component (e.g., hardware, firmware) must be verifiable. Particularly, unauthorized cloning of a protected vehicular hardware component must be infeasible or at least detectable as falsified.

Confidentiality. Unauthorized access to protected or private resources (e.g., trade secrets, personal or proprietary data) must be infeasible.

Integrity. Unauthorized modifications of a vehicular information or a vehicular component must be infeasible or at least detectable either internally or by a regular and non-predictable controlling entity. Particularly, any kinds of replay attacks must be infeasible.

	Vehicle Owner	ECU IP Owner	Flashing Operator	Vehicle OEM
Authenticity	■	■	□	■
Confidentiality	□	■	□	□
Integrity	■	■	■	■
Policy Enforcement	□	□	□	■
Availability	■	□	■	■

Table 6.1: Exemplary mandatory (■) and optional (□) security objectives on flash data of a safety-critical ECU depending on the corresponding entity.

Policy enforcement. Circumvention of effective security policies [1], which all legitimately involved parties have accepted, must be infeasible.

Availability. Authorized entities (e.g., hardware modules, software processes, users) must have proper and timely access to their data and services.

A further, but overall security related objective is *privacy* such that the usage of any vehicular application must not endanger the privacy policies of the respective driver/owner (cf. Section 4.9). However, privacy protection is mainly based on the overall design of a vehicular application, the information actually involved, the specific context, as well as general organizational and legal measures. Security objectives such confidentiality, however, can be necessary to realize privacy protection. Moreover, the concrete security objectives generally depend on the respective application and may be different for each information/resource and each entity involved. As shown in Table 6.1, in the exemplary case of a safety-critical ECU software update, the only entity that may be actually interested in information confidentiality could be the ECU IP owner in order to protect its intellectual property (IP). However—in contrast to the vehicle owner, the flashing operator, and the OEM—the IP owner may be less interested in the OEM's policy enforcement or local availability, whereas each of the involved entities is interested in the integrity of the flash-memory data.

In order to derive the respective security objectives for a security-critical application, the following procedure is proposed. This procedure assumes that a detailed application description and corresponding application use cases are available, but without considering any security issues. Each of following steps is further

---

[1] Such a security policy could be, for instance, an access or execution control policy or a policy that certain hardware components have to be verified for OEM's legitimacy before installation.

exemplarily illustrated by the security objective analysis of a vehicular event data recorder (EDR) [2] application. Anyhow, further readings on this subject are helpful and can be found amongst others in [And01, ISO04a, ISO04b, ISO07b].

(1) Identify all involved, potentially critical data.

(2) Identify all involved entities.

(3) Identify the security objectives of each involved entity on each of the identified data.

(4) Merge the security objectives of all involved entities for each data.

After choosing a security-critical application[3] for SO analysis, all potentially critical data involved are identified. As pointed out by the example of a suddenly uncontrolled spattering windshield washer, even data that seem non-critical at the first glance should be included at the begin[4] of the analysis. In case of the EDR example, such data could be several vehicle sensor data (e.g., current speed or acceleration), vehicle control data (e.g., current steering or pedal positions), the current time, and the current (absolute) position. Now, all involved entities will be identified, that means all concrete or abstract roles and parties, which may have or would like to have access to the data identified before. In the EDR example, this could be the actual driver, the owner, the garage personnel, a supplier, the OEM, several authorities (e.g., police, justice), or for instance an insurance company. Having figured out all involved data and relevant entities, the security objectives of each entity on each of the involved data are identified. In the EDR example, the driver could demand for integrity on the clock data, whereas certain authorities would moreover require authenticity and availability of the clock signal. Otherwise, in contrast to a potentially involved insurance company, the OEM could have no security objectives on actual position and time, but would require authenticity, integrity, and availability for all involved sensor and control data. Finally, to meet the security objectives of all involved entities, the security objectives on each data are merged together. In case of the EDR example, this could result in the combined

---

[2]The event data recorder is an at least tamper-evident device that is already installed in some of today's automobiles and trucks to record vehicle's recent driving-related activities. Similar to a "black box" known from airplanes, information collected from such a device can help for instance to clarify the circumstances of an accident.

[3]Note, as exemplarily shown in the motivating section, even vehicular applications that seem non-critical at the first glance may be susceptible to security issues that can lead to serious impacts on driving safety.

[4]If it turns out afterwards that for certain data none of the involved entities has any security objective on, they are automatically ceased from further analysis.

security objectives authenticity, integrity, and availability on the processed clock data.

6.2 Security Requirements Engineering

Security requirements (SR) are the actual measures or functionalities needed to fulfill the security objectives (SO) identified before. The security requirements, in turn, heavily depend on the actual security environment, that is, where and how the respective application is applied, which assumptions can be made, which security policies are relevant, and particularly, which attack potentials arise. Having identified the effective security objectives, the following procedure is proposed. To illustrate each step of the security requirements engineering as well, the EDR example application from the previous section will be used further on. However, again further readings are helpful and can be found amongst others in [And01, ISO05a, ISO04b, ISO07b, KMS06, MN03, Rus01].

(1) Identify all vehicular components that handle data covered by security objectives.

(2) Identify the effective security environment; concretely, identify effective assumptions, all relevant security policies (if any) and all potential threats (i.e., attacker model, attack vectors[5] or so called "abuse cases").

(3) Estimate the respective attack potentials [CCD07] and application-specific consequences for a potential security breach in order to asses the respective security risks.

(4) Derive and prioritize appropriate security requirements to meet the security objectives for the concrete security environment.

First, all vehicular components that handle (e.g., read, write, modify, or transmit) data, which are covered by security objectives, are identified. This includes components of the actual application as well as supplementary components, which the application shares with other applications (e.g., certain sensors). In case of the EDR example, the involved components could be several sensor and control ECUs, the clock component(s), the internal vehicular communication system (e.g., the CAN bus, FlexRay), and the actual data recorder component. To identify the effective security environment, first potential attackers together with their respective

---

[5]An attack vector is a path, procedures, or means by which a malicious entity can gain a malicious outcome without interpretation of their feasibility.

possible intentions have to be identified. This means a detailed analysis about who will be interested in accessing, destroying or manipulating data (or functionality), that is, who will be interested in circumventing a given security objective and how a potential attacker can gain a malicious outcome (cf. [WWW07]). In the EDR example, a potential attacker could be the current driver trying to manipulate a record, which could be interpreted to his disadvantage. Another attacker could be the OEM or an insurance company, trying to gain more information than they are entitled to (e.g., in order to detect a possible driver's malpractice) that would clearly affect driver's privacy objective. A further part of the security environment analysis is to identify all relevant security policies and effective assumptions. An assumption for the EDR example could be that the actual recorder component is sufficiently physically secure[6]. Effective assumptions also consider feasible access perimeters of potential attackers (e.g., only logical or also physical access), their technical, financial, and knowledge resources and the potential feasibility of an offline attack[7]. In case of the EDR example, the driver may have physical and logical access to the internal vehicular bus system that communicates security-critical data, whereas an insurance company would have only indirect, logical access to security-critical data, but may be able to mount an offline attack.

Having identified all involved components, the effective security environment, and all potential threats, the respective attack potentials can be estimated accordingly, for instance, using the Common Criteria taxonomy [CCD07]. In order to assess also the corresponding security risk[8] for each attack, the different (normally very application-specific) consequences for a successful security breach have to estimated too. Hence, the severity level of a successful security breach can be for instance a function of functional, financial, legal, privacy, or safety consequences. Based on a prioritized list of identified security risks, appropriate security requirements can be derived accordingly, which fulfill the defined security objectives and reduce the corresponding security risks. In the exemplary EDR application, such a resulting security requirement could be *component authentication* that allows the recorder component to verify the identity (and thus the associated assumptions on correctness and trustworthiness) of the clock component, which has to withstand an attacker with basic attack potential. Another security requirement could be *secure bus communication* to prevent manipulations of the clock signal during

---

[6]However, such an assumption should be in turn the result of another independent security analysis.

[7]Attacks on vehicular IT systems are normally *offline attacks*, where an attacker has virtually unlimited time and virtually unlimited trials to successfully mount an attack. Hence, the attacker can calmly mount almost any feasible attack without having to fear to be detected, back traced, or locked out.

[8]The security risk is the product of the probability of a successful attack (i.e., the attack potential) with its potential severity of consequences.

	Actuator	**Driver**	**Garage**	**GPS**	**OEM**	**Police**	**Sensor**
Sensor data	□	r	□	□	r	r	w
Steering data	w	r	□	□	r	r	□
Time, location	□	r	□	w	□	r	□
Service data	□	r	rw	□	r	r	□

Table 6.2: Exemplary discretionary access control matrix for data processed by the data recorder component of an EDR application (r: read access, w: write access, □: no access).

the transmission between the clock component and the actual recorder component, which has to withstand an attacker with moderate attack potential. In the following, some further security requirements, which are typically deduced for security-critical applications in the automotive domain, are given.

Access control mechanisms (e.g., discretionary access control and/or mandatory access control [DoD85]) prevent unauthorized access to restricted vehicular data or restricted vehicular resources (e.g., networks, computing power). Access rules to restricted data and resources are defined in the corresponding security policy derived during the overall security requirements engineering process, which determines the access rights for each authorized entity (cf. Table 6.2).

Component identification & authentication provides verification of the component identifier and the component authenticity. This allows protection of original and legitimate components against counterfeits, thefts, or unauthorized installations by binding critical components to the corresponding vehicle.

Identity Management ensures secure and privacy-preserving creation, assignment, description, management, and deletion of identifiers for users, hardware, software, or processes.

Secure audit protects monitoring of certain vehicular information, actions, and events upon acceptance of all legitimately involved parties. A secure audit then ensures authenticity, availability, and integrity of records.

Secure communication provides confidentiality, integrity, and non-repudiation of the communicated information. It further enables the verification of the authenticity of the communication endpoints (*secure channel*) and could additionally also enable the verification of the configuration of the communication endpoints (*trusted channel*) in order to determine its trustworthiness [AES+07].

Secure initialization ensures the integrity (authenticity, non-repudiation, and freshness) of a vehicular (sub-)system during start of operation as result of a foregoing deactivation or as part of its initial installation.

Secure storage provides confidentiality, integrity, freshness, and availability of information persistently stored.

Secure provision of sensor data provides availability, authenticity, and integrity of information provided by a sensor.

Strong isolation ensures that subsystems, components, and even individual applications can communicate only via strictly controlled communication channels[9] such that it is impossible to access (i.e., data, functionality) or even affect (e.g., performance) each other without proper authorization.

User identification & authentication provides verification of the user identifier and the user authenticity. This prevents unauthorized access to and user-based access control for restricted vehicular data or restricted vehicular resources.

6.3 Characteristical Advantages

By being located between the world of general-purpose computers and the strict embedded world (i.e., cellular phones or smartcards), a vehicular IT environment provides some helpful characteristical advantages, which may considerably ease the implementation of certain security requirements. Some of them are briefly described in the following.

Feasibility of updates. Even though software and particularly hardware updates are quite restricted in extent and frequency, they are at least feasible in a limited manner. Particularly, since automotive OEMs are usually liable for up to 20 years if critical faults or vulnerabilities have been identified. Thus, they are heavily interested in timely applying necessary (security) updates. Vehicle owners are usually interested in an up-to-date status just as well to ensure reliability and property retention of their vehicle.

Periodic inspections. Vehicles are usually subject to periodic predictable (e.g., by a technical inspection authority) and non-predictable (e.g., by the police) inspections by an official control entity so that possible (successful or even attempted) attacks often can be detected afterwards and lead to non-technical (legal) actions.

Moving target. Since a vehicle usually continuously changes its physical location (i.e., in contrast to a general-purpose computer system), at least an external attacker has comparatively limited time and limited trials to successfully attack and encroach a vehicle.

---

[9]Strictly controlled communication channels can be provided by a very small and hence verifiable hardware-based and/or software-based separation mechanism such as ARM's TrustZone technology [WFM+07] or several virtualization technologies [MLO97, SJV+05].

Physical protection. Even though many security-critical vehicular applications have to deal with attackers, which may have also physical access to crucial components, vehicles usually provide, up to a certain extent, some physical protection that, according to the attacker's access perimeters, may at least complicate many attacks considerably.

Sufficient energy and space. In contrast to strictly embedded devices (e.g., mobile devices or smartcards); there are somewhat weaker restrictions on power consumption and devices' size and weight, which allow implementing somewhat more sophisticated and costly (with regard to size and power) security functionalities.

Ongoing standardization efforts. Ongoing standardization of involved (security) hardware, software, interfaces and protocols, indeed makes potential attacks more scalable, but, on the other hand, also allows regular careful verifications for correctness and immunity, makes security less costly and reduces the application of error-prone proprietary security solutions.

Centralized controller topology. Necessary functional modularity [Lar05] previously was realized by many individual[10] functionally independent hardware controllers that could easily be added to or removed from the vehicular IT system. However, since this traditional approach turned out to become increasingly complex, costly and hardly controllable, current approaches employ only three or five centralized powerful controllers [CCA02, Fri04] that merge the functionality of several individual ECUs. Such a centralized controller topology can considerably ease the implementation and maintenance of security measures. A few powerful, flexible, centralized vehicular controllers enable powerful, flexible, and sophisticated security measures, which in turn can reliably protect a large set of vehicular functionalities.

6.4 Characteristical Constraints

Despite some helpful advantages (cf. Section 6.3), vehicular IT environments also implicate some characteristical technical and non-technical constraints, which may considerably affect, restrict, or even prevent the realization of certain security requirements. Some of them that have to be carefully considered are briefly described in the following.

---

[10]In today's premium vehicles, one can find up to 80 individual ECUs [Bro06].

6.4.1 Technical Constraints

In the following several typical technical constraints, which may affect the realization of various vehicular security requirements, are described.

Limited computing resources. Computing resources of vehicular components are, in comparison to general-purpose computer systems, rather limited due to the typically strong cost (weight and energy) requirements. Nevertheless, automotive applications are often required to provide (hard) real-time capabilities. This leads to severe restrictions on complexity, memory size, and runtime efficiency for automotive security implementations that moreover often have to cope with lots of specific architectural restrictions, which often means costly, low-level, and hardware-specific implementations. However, highly customized code increases maintenance and decreases its reusability.

Physically challenging environment. Vehicular IT systems are often subject to specific physical constraints such as high variations in temperature, moisture or particular mechanical loads. They have to cope with these conditions usually over a product life cycle of up to 20 years in which only minimal maintenance efforts are acceptable.

Limited external communication resources. A vehicular IT system usually has only very limited communication resources to, for instance, exchange cryptographic keys, update certificates, or to adapt revocation lists. Thus, virtually all vehicular security functionality has to work properly even with an external communication functionality severely limited in capacity and frequency.

Limited update feasibility. The application of frequent security updates is, in comparison to general-purpose computer systems, rather limited. Even if software-based security issues could be fixed almost automatically using available external communication resources, security vulnerabilities in hardware components usually would require costly and irritating recall procedures that means the vehicle owner has to locate for an authorized garage to fix up the respective component. Moreover, once millions of vehicles are sold, over many years, all over the world, it is almost infeasible to fix a critical vulnerability afterwards on all vehicles in use.

Limited complexity for user interaction. Since typical computer users can mostly employ ergonomic input and output devices to accomplish for instance user authentication or certain security settings, users within the automotive environment are restricted to only little ergonomically designed man-machine interfaces (MMI). To demand only a minimum of user interactions, virtually all vehicular applications are required to run almost completely autonomous. However, if user interactions are inevitable, they have to be practicable as simple, flexible and little cumbersome as possible.

Increasing system complexity and diversification. The high complexity as well as the multifunctional, diversified, and distributional nature of current vehicular electronics is contrary to the security principle of economy, such that corresponding security measures can be "designed to be as simple as possible, so that the mechanism can be correctly implemented and so that it can be verified that the operation of the mechanism enforces the containing system's security policy" [Shi00]. Current vehicles already provide up to 2000 individual software-based functions, which inherently generate several ten thousand pages of technical documentation [Bro06]. Future vehicles will further increase application diversity. This clearly complicates a comprehensive realization, and in particular a complete verification, of security measures that are able to reliably protect such a complex IT system. However, ongoing standardization efforts (cf. Section 6.3) will at least help to counteract an excessive proliferation of proprietary functionalities.

Distributed architecture. Vehicular IT systems and the involved components (e.g., sensors, busses, or controllers) usually are very heterogeneous and widely distributed over the whole automobile. This creates many vulnerable points to successfully mount an attack, and hence considerably complicates a holistic protection of security-critical applications.

6.4.2 Non-Technical Constraints

Beyond the characteristical technical constraints, automotive IT systems are also subject to several characteristical historical, organizational, and legal constraints, which may considerably affect or complicate the realization of various security requirements.

Unfamiliar technology. IT security was too long an only little-noticed subject in the development of vehicular applications and services. With the rapid introduction of more and more software-driven vehicular components, IT security suddenly becomes an essential technology for many automotive developments. However, the characteristical traditional structures in the automotive industries, with their multitude of proprietary and isolated developments, make it rather difficult to properly implement IT security, which particularly demands holistic, top-down approaches. Strictly speaking, current vehicular infrastructures were never intended to implement also IT security measures. Thus, there hardly exist any up-to-date standards, rules, or specifications regarding vehicular IT security, while current standardization and specification efforts can hardly keep up with the fast ongoing development of vehicular electronics. Furthermore, automotive engineers normally are mechanical engineers or electrical engineers without any special training in security. Given the tricky pitfalls often inherent in security design, this is often a major real-world hurdle[11]. Hence, the automotive industry is missing experts in IT security, which are already rare in most other industries with a need for security experts (e.g., aircraft industry).

Isolated subsystem development. For historical reasons, most vehicular subsystems are developed and produced completely independent from the corresponding OEM by individual independent suppliers, which in turn often serve several different OEMs at the same time. However, current software-driven development considerably increases mutual relation and interaction of vehicular subsystems with each other, which makes independent and autonomous security measures often impractical or even impossible. Holistic security measures generally require integrated top-level approaches; that is, OEMs have to synchronize the developments of their suppliers and thus have to jointly agree on developments for subsystems related with each other to solve dependencies and prevent vulnerable interactions.

Multitude of involved parties. The current multi-tier vehicular manufacturing process (OEM and possibly several layers of suppliers) can complicate the realization of security solutions considerably. It will be difficult, for instance, to decide who is in charge for the overall security design and, in particular, who has control over the corresponding cryptographic secrets. A potential public key and certificates infrastructure (PKI) as well as the overall security management requires further complex and costly organizational structures and involves even more different parties (e.g., manufacturer, supplier, OEM, garage personnel, content provider, etc.) with very different security understanding (cf. Chapter 9).

Additional costs with little promotional benefits. In contrast to the implementation of novel functionalities, implementing IT security first of all creates costs without apparent (i.e., promotional) benefit for suppliers, OEMs, and customers. Since it is even hardly possible to estimate corresponding benefits as a result of prevented attacks, it is yet difficult to make a useful internal cost-benefit calculation.

Long product life cycles. Since vehicular IT systems—in comparison to, for example, usual operating system software—have only limited possibilities for maintenance; simplicity, stability, and reliability of deployed hardware and software are obligatory requirements. Compatibility, portability, and reusability are further important requirements for all implemented security solutions, as it is not unlikely

---

[11] As it for instance happened with the DVD video copy protection [Pat99], an ill-conceived security design is quickly broken.

that a vehicular manufacturer runs out of certain microelectronics due to the typically rather short semiconductor product life cycles. Moreover, all security mechanisms and corresponding infrastructures have to be designed for proper operation during the complete life cycle of a typical vehicle (e.g., by flexible parameters, usage of long-term security values, foreseeing of upgrades), that means up to two decades.

Comprehensive liability. As vehicular IT is often involved in highly safety-critical applications (e.g., driving assistances, crash prevention), they cannot released "without any warranty" and "at owner's risk" as most general-purpose computer software usually does. To provide operating safety and legal security, legally binding warranties are mandatory solely due to legal liability and government legislation. However, warranty statements can usually be given only based on complex and expensive internal and external certification procedures [Ame04]. Thus, corresponding documentation, models, tests, and assessments as well as the development process itself have to be prepared for possible certifications already at the beginning of every development process.

Interoperability and compatibility. An important key factor for most IT security solutions is interoperability to existing (security) infrastructures and devices to enable end-users to integrate their existing devices (e.g., mobile navigation systems, smart phones, multimedia players) as simple and holistic as possible. Closely related to interoperability is compatibility, which assures that updated or upgraded vehicular (security) functionality does not harm or lower other existing (security) functionality. However, due to the increasing complexity, functional dependencies, and mutual interactions (cf. Section 6.4.1), the verification of compatibility becomes even more complicated.

Usability and limited willingness for user interaction. Any security measure becomes worthless if it is not (correctly) applied by the intended entities due to a bad or cumbersome usability. Regularly changing passwords for instance, would potentially increase the security of an authentication system, but often yield to quite the contrary. Since it is difficult for people to choose and remember a new password every month, they usually begin to choose much weaker "serial passwords" such as number series or month names. Moreover, many security mechanisms come with several inconveniences for the involved users (without a directly apparent benefit). However, the willingness of users to spend time and effort on a security mechanism that is difficult to understand, time-consuming or error-prone is very limited.

Various patents and regulations. IT security is subject to various patents, many different cryptographic laws, protection claims and regulations (e.g., import/export regulations [Koo08]), which can complicate the overall congeneric deployment of IT security measures required by the global automotive market.

Limited influence on microelectronic developments. Even though current vehicles have already built-in a huge amount of microelectronic components, the automotive industry currently represents (cf. Figure 6.1) at most [12] about 15% of the microelectronic world market [STM07]. Hence, the influence of the automotive industry on the developments of the semiconductor industry is rather limited to encourage, for instance, the development of customized security solutions specially adapted for the typical requirements in the automotive domain.

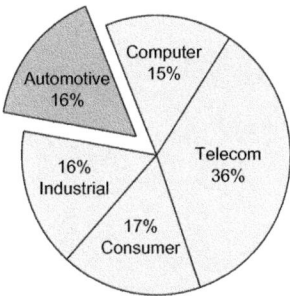

Figure 6.1: STMicroelectronics revenues per market segment for the second quarter sales in 2007 (Q2/07) with a total value of $2.42 billion US dollar [STM07].

---

[12]STMicroelectronics is the largest European semiconductor supplier and one of world's largest supplier of automotive microelectronics.

Part III

The Protection

7 Vehicular Security Technologies

This chapter provides an overview about general vehicular security technologies such as physical security measures, vehicular security modules, and vehicular security architectures. These technologies serve as basis to implement identified security requirements using the security mechanisms described in the next chapter. Parts of this chapter are based on published research in [BEPW07, BEWW07, HSW06, SSW06].

7.1 Physical Security

In contrast to most other IT related attack scenarios, attackers in the automotive domain usually have full physical access to breach the security of a particular vehicular IT system. As described in detail in Section 5.2.2 about physical attacks, an internal attacker in the automotive domain can manipulate or replace almost every built-in component and can manipulate its actual physical environment and (physical) inputs. He further holds the respective attack target in his possession for as long as he likes, and may eventually even receive more samples for testing and practice. Hence, the attacker can undisturbedly mount almost any feasible attack without having to fear to be detected, backtracked, or locked out. Nevertheless, there exist several measures to make physical attacks at least more difficult, even though it is practically impossible to fend off a sufficiently motivated (and sufficiently funded) attacker completely. Thus, a security-critical IT system cannot solely rely on its physical protection measures and hence has to ensure that the successful compromise of a single hardware component does not compromise the overall IT system. This means that the cost of compromising a single hardware component should generally outweigh the potential rewards (*economic security*).

Physical security or *tamper protection* measures usually either aim to prevent any kind of disclosure and modification (*tamper-resistance*), or aim to at least enable a subsequent detection of potential disclosures or modifications by a regular and unpredictable examining control entity (*tamper-evidence*). Physical security measures can be further distinguished into active (*tamper-responsive*) and passive (*tamper-evident, tamper-resistant*) protection measures. This results in the following three definitions.

- Being *tamper-evident* refers to a passive physical security characteristic, which provides detection whether a hardware component has been illicitly modified or compromised. Optionally, tamper-evidence provides moreover the detection of unsuccessful tampering attempts. However, tamper-evidence itself cannot prevent any potential modifications or disclosures.

- Being *tamper-resistant* or *tamper-proof* refers to a passive physical security characteristic, which prevents an attacker from illicitly modifying or compromising a hardware component by passive, non-responsive physical protection measures.

- Lastly, being *tamper-responsive* refers to an active physical security characteristic, which actively prevents an attacker from tampering a hardware component by triggering appropriate counteractive measures up to automatic self-destruction. Tamper-response, in turn, is based on tamper-detection measures, which have to detect an ongoing attack in order to trigger proper response measures.

However, deploying physical security measures at the same time means that the maintainability of such a protected hardware component usually will become clearly limited. This holds, since it is normally impossible that a tamper-protection measure is able to distinguish between an authorized access and an unauthorized access. In many cases, physical security can be increased by applying tamper-evident, tamper-resistant and tamper-responsive measures in parallel, or by applying them in a layered manner, where particularly critical sub-components inside of a protected hardware component have furthermore an additional individual tamper-protection with yet a higher protection level. However, concrete physical security measures are closely connected with various application-depending details and the individual security objectives respectively. Hence, in the following firstly a basic overview of each of the afore distinguished physical security measures is given. Also, further readings are helpful and can be found amongst others in [FIP02a, ISO07a, KK99, Lem06, Wei00].

7.1.1 Tamper-Evidence

Tamper-evident physical security measures require a regular and non-predictable controlling entity, so that potential attacks or even potential attackers can be identified at least afterwards to take appropriate non-technical actions (e.g., legal actions). Thus, manipulations on digital tachographs, for instance, cannot be prevented, but can be detected by a respective official that then leads to corresponding legal actions. Common tamper-evident security measures in the automotive domain use official seals, special labels (e.g., holographic stickers), or special packages (e.g., brittle packages, crazed aluminum, polished packages, bleeding paint). Thus, removal, replacement, or counterfeiting of tamper-evident seals, closures, or packages should be infeasible or should at least leave sufficient marks that are detectable afterwards by the controlling entity. Nevertheless, as described amongst others in [JG97], most of the currently deployed tamper-evident measures can be defeated already by using only publicly available supplies and quite simple technical equipment.

7.1.2 Tamper-Resistance

Tamper-resistant protection measures cannot detect the tampering itself (directly), but try to prevent any physical intrusions or try to make them at least as difficult as possible. For that, it uses special packages (e.g., hardened steel, susceptible sealing, epoxy coating, ceramics), special interlockings (e.g., security screws, welding), very small semiconductor structures (e.g., 90 nm technology or less), shielding mechanisms, or logical measures such as bus and memory encryption, or sophisticated methods of obfuscation.

Attacking these preventive protection measures, however, often also provides tamper-evidence as a side effect. Tamper-resistance further tries also to prevent or complicate non-invasive attacks that apply external probing techniques such as side channel attacks (cf. Section 5.2.2) or optical probing, which do not harm the attacked hardware physically. This can be achieved for instance by using materials that are impermeable for the critical electromagnetic frequencies (e.g., visible spectrum, infrared, microwaves), shielding mechanisms, or special semiconductor designs (e.g., obfuscating multi-layered structures, dual-rail logics). However, particularly side channel protection usually furthermore requires algorithmic or logical countermeasures within the actual implementation of the respective functionality such as data masking or dummy cycles [SLP06] to reveal as little information as possible by physical side effects.

Lastly, tamper-resistance may also include some self tests that were regularly executed. Hence, for instance on every power-up, the integrity of all critical internal information and resources (e.g., the random number generator) are verified accordingly. Thus, a potential tampering attempt can be detected at least afterwards to be thwarted somehow (cf. *tamper-response*).

7.1.3 Tamper-Detection

Tamper-detection mechanisms are required to sense an ongoing attack before triggering appropriate response measures. This includes sensing active intrusions as well as sensing abnormal operating conditions (e.g., out of range temperatures). Intrusion detection can be based for instance on electromechanical sensors that recognize a disassembly (e.g., microswitches, pressure contacts, motion sensors, light detectors) or sensors that monitor intrusion detection meshes (e.g., wire meshes, piezo-electric sheets, fiber optics) wrapped around critical hardware areas that recognize small changes of the mesh's capacitance or resistance.

Environmental sensors can detect out of range values or abnormal changes for temperature, voltage, radiation (e.g., X-rays, ion beams), light, or clock frequency to recognize critical operation conditions that suggest a physical tampering for instance by launching fault attacks or glitch attacks (cf. Section 5.2.2). To assure tamper-detection mechanisms also against offline attacks (cf. Section 5.2), the protected hardware component normally requires an autonomous internal power supply.

7.1.4 Tamper-Response

Once a physical tampering has been detected (cf. tamper-detection), tamper-responsive measures actively try to protect the internal assets by triggering appropriate countermeasures. Tamper-responsive measures therefore can range from the interruption of the security-critical operations or their complete deactivation, the deletion of internal secrets, up to the physical self-destruction of the respective hardware component (e.g., by a small explosive charge). Currently, most tamper-responsive measures try to immediately erase internal secrets by RAM power drop or RAM overwrite (*zeroization*) and very seldom by automatic physical destruction. Even though tamper-responsive measures ideally operate without having a internal power supply available, they normally require an internal autonomous energy source such as very small, long-life batteries.

7.2 Security Modules

A *security module*, which is sometimes also called a *security anchor*, usually means a dedicated cost-effective cryptographic microcontroller securely attached to the vehicular hardware that has to be protected. The security module itself, in turn, is specially protected by physical security measures, which ensure tamper-resistance or at least tamper-evidence (cf. Section 7.1). It securely provides general security-critical functionalities such as cryptographic operations (e.g., encryption, decryption, or hashing), secure timing or secure random number generation to enable various higher security mechanisms such as secure initialization, secure communication, or secure storage. By employing a hardware-protected security module, security-critical information (e.g., secret keys) and corresponding computations are securely protected by hardware measures and hence cannot affected by any (malicious) software functionality.

However, many security modules have to rely on susceptible external resources, for example, for power supply or external clock signals. Hence, highly critical security modules have their own battery, next to their own clock signal generator and their own independent random number generator. On the other hand, as already mentioned in Section 7.1, it is practically impossible to fend off a sufficiently motivated (and sufficiently funded) attacker completely. Thus, a security-critical vehicular IT application cannot solely rely on the tamper-protection of its security modules, and hence has to ensure that the successful compromise of a single security module does not compromise the corresponding IT application for all vehicles in the field. Briefly summarized, a proper security module has to realize the following basic security requirements.

Secure creation. Unauthorized production of identical copies (i.e., cloning) of the security module must be infeasible.

Secure integration. Unauthorized installation of the security module into another vehicle must be infeasible. Hence, the security module must be securely bound to the particular identity of the corresponding vehicle such that it cannot be installed into another vehicle without proper authorization.

Secure operation. Affecting internal operations of the security module (e.g., by fault attacks) or leakage of secret internals as a result of its internal operations (e.g., by applying side-channel attacks) must be infeasible.

Secure storage. Unauthorized readouts, manipulations, or deletions of internal storage must be infeasible.

Secure output channel. The security module has an authentic integrity-preserving output channel to securely provide critical or protected information or to be able to raise an alarm if necessary.

While the requirements on secure creation (i.e., counterfeit protection) and secure integration are subject to component identification mechanisms (cf. Section 8.2), the requirements for secure storage, secure operation, and the secure

output channel are subject to physical protection (cf. Section 7.1) in combination
with appropriate cryptographic measures (cf. Section 8.5).

	Software module	Std security controller	TC security controller	FPGA security box	ASIC security box
Standardized	Possible	Yes	Yes	No	No
Flexibility	Maximum	Very limited	Very limited	Full also after release	Full until release
Performance	Moderate	Adaptable	Low	Fast	Maximum
Expenses	Very low	Varies	Moderate	High	Varies
Security level	Very low	Varies	Medium	Can be high	Adaptable
Internal attacks	Feasible	Demanding	Feasible	Preventable	Preventable
External attacks	Preventable	Preventable	Preventable	Preventable	Preventable

Table 7.1: Comparison overview of various vehicular security module realizations.

A security module can be realized as a pure software module, or by using a
standard security controller, a Trusted Platform Module (TPM) as proposed by the
TCG [Tru03], or a customized security box based on a particular ASIC or based
on a flexible FPGA processor. However, each possible solution has its individual
advantages and drawbacks, which are firstly summarized in Table 7.1 and then
individually described in the following subsections.

7.2.1 Software Module

The easiest way to realize a security module is to implement it purely in software. Also, this is usually the most flexible and inexpensive solution. However,
the realization of a security module as a pure software application is, at least in
resource-restricted vehicular environments, normally in order of magnitudes less
powerful than a dedicated hardware-based approach. It further provides a considerably lower security level than most hardware-based solutions, since pure software security mechanisms in a hostile environment can often be broken very easily [HM04, vO03]. However, by carefully regarding the principles for secure software development and implementation (cf. Section 8.4 et sqq.) together with a
strict review and verification process[1], at least runtime or online attacks that try
to exploit available software interfaces can usually be thwarted effectively. However, in the automotive domain most encroachments are based on hardware mod-

---

[1] Ideally, the software security module should be evaluated formally using an internationally accepted
security evaluation standard such as the Common Criteria [ISO05a].

ifications such as memory readouts, program memory manipulations, or communication line probing, which cannot be prevented (or even detected) by a purely software-based security module.

7.2.2 Security Controller

A security controller refers to an enhanced microcontroller as normally employed in smartcards. Thus, a security controller usually is protected against most physical attacks[2] and provides basic cryptographic functionality for symmetric and asymmetric cryptography as well as cryptographic hash functions. It usually also provides a protected random number generator and a small amount of secure storage. There exist dozens of models from cheap, small and simple 8-bit controllers up to costly, high-performance 32-bit controllers with a large amount of memory and a versatile functionality. However, once a particular security controller model has been chosen, it is hardly possible to change, update, or upgrade its functionality afterwards.

Due to their mass production characteristics, security controllers usually are relatively cheap, adequately standardized, and their security level usually well classifiable based on generally accepted security evaluation methods such as the Common Criteria [ISO05a]. Carefully implemented and thoroughly evaluated, a security controller is able to reliably thwart all external non-physical attacks. Further, even though its physical security concretely depends on its very individual tamper-protection capabilities, a usual security controller provides adequate protection against internal attacks, which includes attacks based on physical access. Nevertheless, standard security controllers often fail to fulfill the tight requirements for cost-efficiency, adaptability, and physical durability that are characteristical in the automotive domain (cf. Section 6.4).

7.2.3 Trusted Platform Module

A Trusted Platform Module (TPM), which currently serves as the base of Trusted Computing technology (cf. Section 3.8), means a tamper-resistant hardware device similar to a smartcard that is assumed to be securely bound to the respective (vehicular) hardware device. As previously described in Section 3.8.1, a TPM implements elementary security functions while keeping the costs as low as possible,

---

[2]Lastly, the individual implementation defines which physical attacks can be actually thwarted and which not.

and thus keeping the assumptions on tamper-evidence as weak as possible[3]. Current TPMs are based on the specifications version 1.2 [Tru07b] published by the Trusted Computing Group (TCG) and currently available for desktop computers and embedded systems. As depicted in Figure 3.8, a TPM basically consists of a hardware-based random number generator (RNG), a cryptographic engine for encryption and signing (RSA) as well as a cryptographic hash function (SHA-1, HMAC), read-only memory (ROM) for firmware and certificates, some volatile memory (RAM) for secure operation, non-volatile memory (EEPROM) for internal keys, monotonic counter values and authorization secrets, and optionally, several sensors for tampering detection (cf. Section 7.1).

As detailed described in Section 3.8.2, a TPM chip provides some basic security functionalities on which a larger set of security mechanisms can be built. Thus, a TPM securely implements a set of cryptographic operations that cannot be compromised by potentially malicious software. It further provides a functionality for hardware-protected device integrity measurements and reporting, which reflect the device's current hardware and software configuration. A further distinctive feature of Trusted Computing hardware is the ability to not only use passwords as authorization for several critical cryptographic operations (e.g., decryption). Moreover, it also allows to use the afore mentioned integrity measurements as effective authorization. That is, only a device running a certain, previously defined hardware and software configuration is authorized to use, for instance, a certain key for a decryption operation. Moreover, the property that a certain key is "bound" to a device configuration can be certified by Trusted Computing hardware. A remote party then can verify the certificate and validate the embedded integrity measurements against "known good" configurations before encrypting data with the certified key.

However, the actual functionality of a TPM is strictly determined by the corresponding TCG specifications [Tru03], and hence quite limited regarding individual changes, updates, or upgrades. A usual TPM further needs considerable software support, due to the inherent limitations in complexity and performance of its cryptographic hardware functionality. Since a TPM, per definition [Tru07b], should thwart software attacks only, it indeed can thwart most external attackers, but it usually cannot thwart internal attackers that have physical access to the TPM chip and its corresponding peripherals. On the other hand, the strict standardization efforts of the TCG enabled a wide range of interoperable security applications and further allows the expansion of many TPM-based applications from the multimedia and mobile world also into the automotive domain. And even though TPMs are, by comparison, currently still somewhat expensive, once they will become

---

[3]Nevertheless, some TPM manufacturers have third party certifications according to Common Criteria's security evaluation assurance level 4+.

deployed on a large scale, TPMs or at least a specially shortened automotive TPM version, which efficiently meets the automotive requirements, could become economically feasible even in the automotive domain.

7.2.4 Security Box

A security box refers to a fully customized security hardware that is usually realized by an application-specific integrated circuit (ASIC) or based on a field programmable gate array (FPGA). The respective functionality, performance, capacity, and physical protection measures can be completely defined individually by the corresponding customer and corresponding application requirements respectively. Security boxes hence can provide high performance values by employing dedicated hardware and strong physical security measures at the same time. On the other hand, such highly customized security devices usually involve considerable development costs and considerable development time. Due to their proprietary nature, security boxes usually involve a costly (i.e., in terms of time and financial expenses) individual certification process and are only limited interoperable and reusable that often conflicts with the highly modular approach for automotive development and production. In conjunction with the generally only little production quantities, customized security boxes were so far reasonable only in military, embassies or intelligence services, and a few very security-critical industries. However, future highly safety and security-critical vehicular applications (cf. Chapter 4), could make the deployment of security boxes reasonable even in the automotive domain [HSW06].

7.3 Vehicular Security Architectures

In the ideal case, a security module can be deployed, according to the respective security objectives, in every vehicular component that has been identified security-critical for a certain security-critical application. However, integrating a fully-fledged security module into every security-critical vehicular component would inherently cause considerable costs and a huge complexity. There are at least three different vehicle architectures to provide security functionalities within a vehicle, without necessarily providing every vehicular component with an individual security module [BEPW07]. The individual characteristics for the *central*, the *distributed*, and the *semi-central* approaches for a vehicular security architecture are firstly summarized in Table 7.2 and then individually described in the following subsections.

	Central Architecture	Distributed Architecture	Semi-central Architecture
Overall security level	Rather low	Can be high	Adaptable
Secure integration of other controllers	Requires further security measures	Fully possible	Partly possible
Expenses	Low	High	Moderate
Complexity	Low	Moderate to high	Moderate
Flexibility	High	Moderate	Moderate

Table 7.2: Comparison overview of the three different approaches for a vehicular security architecture.

7.3.1 Central Security Architecture

In a central security architecture, a single central security module provides the security functionality for all other internal and external vehicular controllers and devices. The central security module can be implemented separately, but is ideally part of the central control unit or head unit that already implements most applications with a need for protection. Since, therefore, solely the central security module provides some reasonable physical protection, it serves as the central internal and external communication point that stores all security-critical information and implements all crucial security applications such as the electronic immobilizer, the digital tachograph, or the software update mechanisms. Moreover, other devices can employ it for their key management and key establishment procedures (e.g., to establish secure channels among each other), for protected storage, or mutual authentication.

However, devices without any own security functionality have to trust the central security module completely, just like the central security module in turn, which cannot verify its counterparts for their correctness. Hence, without any additional security measures (e.g., a physically protected communication), unprotected devices can neither check or protect any information they receive, nor can they assure (and be trusted for) the correctness of the information they provide to others. Hence, only devices or controllers that implement some own security functionality, at least for secure authentication (e.g., based on a shared secret), can be employed for security-critical applications (cf. with *semi-central security architecture* described afterwards). The central security module, in turn, has to rely on other (non-technical) mechanisms (e.g., trustworthy mechanics) that assess or verify the correctness and hence the trustworthiness of other devices before employing them in a certain security-critical application. Thus, a central security architecture based on a single central security module is:

- easy to realize and easy manageable,

- quite inexpensive, and

- flexible and upgradeable.

But on the other hand:

☐ information from other devices cannot be verified for correctness,

☐ security objectives on information to other devices cannot be enforced,

☐ a successful attack on the central security module defeats the whole vehicular security architecture (*single point of failure*).

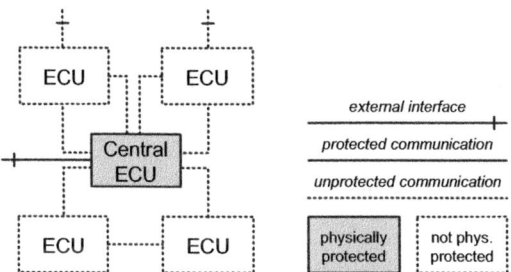

Figure 7.1: Exemplary central vehicular security architecture, which employs a single protected central security module that provides the security functionality for all other internal and external vehicular controllers and devices.

7.3.2 Distributed Security Architecture

In a distributed vehicular security architecture, all involved devices and controllers implement parts of the overall vehicular security functionality. Hence, a distributed security architecture can be either based (i) on several autonomous fully-fledged and fully-protected security modules, or (ii) on a mandatory collaboration of multiple vehicular devices, so that a certain subset of devices and controllers is required to collaborate with each other in order to enable a certain security functionality.

Thus, necessary security functionalities are provided either individually by autonomous and individually protected security modules, or by an abstract security module formed by the mandatory collaboration of a subset of devices and controllers. Thus, a distributed security architecture is:

■ hard to attack (since it requires several individual attacks), while

■ information from other devices can be verified for correctness,

■ security objectives on information to other devices can be enforced, and

■ a successful attack on a single security module does not defeat the whole vehicular security architecture.

But on the other hand, a distributed security architecture is:

□ hard to realize in practice, since it either:

  □ requires sophisticated cryptographic schemes (e.g., Shamir's secret sharing [Sha79]) that are currently not feasible in practice, or

  □ requires expensive and complex individually implemented security functionality and hence individually protected devices,

□ moderately flexible, and

□ moderately upgradeable.

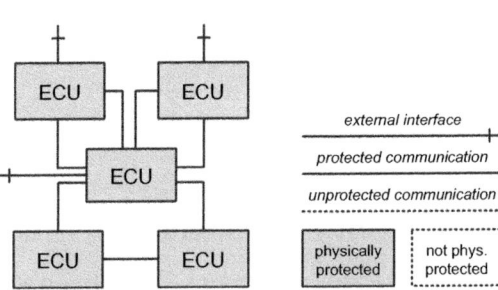

Figure 7.2: Exemplary distributed vehicular security architecture, where all involved vehicular devices and controllers are individually protected and autonomously implement some own security functionality.

7.3.3 Semi-Central Security Architecture

The semi-central security architecture is a combination of the central security architecture and the distributed security architecture. For this, it employs a fully-fledged central security module, but additionally employs some individually secured vehicular devices and controllers (e.g., crucial sensors or actuators) that implement some autonomous security functionality to become securely integrated into security-critical applications. Autonomously implemented security functionalities are usually related to secure authentication, secure communication, or secure storage. Thus, with the help of the fully-fledged powerful central security module, the other controllers, in this architecture, have to implement only a subset of the actually required security functionality (cf. *distributed security architecture*), but are able to securely utilize the extended security functionalities provided by the central security module (cf. with *central security architecture*), which they do not have implemented or for which they have not sufficient resources. Thus, a semi-central security architecture is:

- practically feasible,

- sufficiently flexible and upgradeable,

- sufficiently hard to attack (since critical controllers are individually protected), while

- information from critical devices can be verified for correctness,

- security objectives on information to critical devices can be enforced,

- a successful attack on a single security module defeats only parts of the vehicular security architecture (no *single point of failure*).

But on the other hand, a semi-central security architecture is:

- ☐ not that simple, flexible, easy manageable as a central security architecture, and hence further

- ☐ moderately expensive and complex.

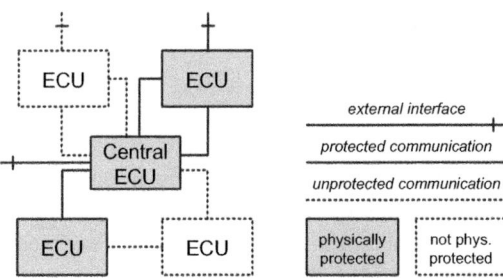

Figure 7.3: Exemplary semi-central vehicular security architecture, which employs a fully-fledged central security module, but additionally employs some individually secured vehicular devices and controllers that securely implement some security functionality autonomously.

8 Vehicular Security Mechanisms

This chapter provides an introduction into feasible security mechanisms applicable in the automotive domain to fulfill the identified security objectives and necessary security requirements. Parts of this chapter are based on published research in [AES⁺07, HPWW05, HSW06, PWW04a, PWW04c, PWW05, SSS⁺06, SSW06, WWW07].

8.1 Why Proper Security Application is Hard

Several security mechanisms and technologies exist to realize the identified vehicular security requirements (cf. Section 6.2). However, just as important as the inherent security level of a certain security mechanism is its proper application, implementation, and interaction with other security mechanisms. As exemplarily depicted in Figure 8.1, even a provable secure mechanism can quickly become worthless if it is applied in an improper and hence insecure manner. Another alerting example are multiple encryptions or the cascading of ciphers, where one could assume that the security of an encryption is automatically increased by combining multiple encryptions. However, cascading multiple encryptions could effectively result in quite the contrary [Sch96]. Thus, any security design should be evaluated by skilled security experts to detect and prevent at least obvious security vulnerabilities and weaknesses due to improper application, improper implementation, or improper interaction as early as possible.

Another important issue for proper application of security mechanisms in the automotive domain is interoperability. Having multiple isolated security developments and implementations usually is costly and inefficient. This may result for instance in a multitude of incompatible security hardware and proprietary software solutions, several incompatible (key) infrastructures, and many emergency or less-than-ideal security solutions. OEM-wide holistic and interoperable security approaches, in contrast, could enable various valuable synergies that would reduce many redundancies in security development, implementations and security evaluations by increasing the reuse of well-proven designs, available infrastructures, and available security hardware and corresponding software. Moreover, if several security applications require the same security mechanisms (e.g., for se-

cure storage), it further could become economically justifiable to employ a higher
security level solution, from which in turn, all corresponding applications would
benefit.

(a) Original image.   (b) ECB encrypted.   (c) CBC encrypted.

Figure 8.1: The original image has been encrypted using a secure symmetric cipher in
electronic code book (ECB) mode and in cipher-block chaining (CBC) mode of operation [Wik08], showing that security depends on more than solely secure mechanisms.

8.2 Secure Component Identification

Dependable component and vehicle identification is a crucial requirement for
many vehicular applications such as electronic license plates or most VC applications (cf. Chapter 4). It is particularly required to preserve vehicular safety
by ensuring the authenticity (i.e., genuineness) of safety-critical vehicular components such as airbags or several driving assistants. Moreover, in addition to the
vehicular safety implications, the financial loss for vehicular manufacturers and
suppliers due to counterfeiting alone is estimated to be as high as several hundred
billion Euros worldwide [PT06].

   Even though physical security measures can help to protect vehicular components against various (physical) manipulations or readouts, tamper-protection measures seldom help to identify faked, stolen, or unauthorizedly [1] assembled vehicular components. Hence, even a perfect tamper-resistant component, for instance an
electronic license plate that securely broadcasts the unique identity of the vehicle,
could just be stolen and installed into another vehicle to accomplish various impersonation attacks (e.g., speeding at another's expenses, broadcasting malicious

---

[1] Unauthorized components refers to components such milage counters or airbag systems that should
not be changed without the approval of corresponding officials.

VC messages). Thus, additional component protection measures are required to securely bind crucial components to the respective vehicle and hence to enable detection that a crucial vehicular component either is missing or is not authentic.

Existing traditional methods for tamper evidence such as holographic stickers, seals, or special labels (cf. Section 7.1) are used to physically link security-critical or safety-critical components to a certain vehicle or, at least, to provide an evidence for their authenticity. Other approaches rely on mechanical countermeasures, which use special proprietary constructions that fit only into vehicles of a particular manufacturer or that require proprietary (i.e., not publicly available) tools and equipment. However, these solutions are often unpractical, uncomfortable, and above all, provide actually only minimal security [JG97].

Nevertheless, vehicular manufacturers are already developing some more sophisticated approaches that use for instance component identity codes. As described in [GHOP01], for instance, unique identity codes are written into a small digital memory of the respective component by the manufacturer according to the specifications of the OEM. During the initial startup of the vehicle, the vehicular operating system then retrieves and verifies all identity codes available. If a crucial component is missing or does not belong to a specific vehicular setup, an error is raised. However, such approaches are usually based on trustworthy mechanics or can easily be circumvented as they lack any cryptographic or physical protection measures. They hence can neither dependably prevent component thefts (i.e., unauthorized removal and installation respectively) nor counterfeits or manipulations. Thus, in the following two more dependable component identification schemes are introduced, which are based on cryptography and physically unclonable functions (PUFs).

8.2.1 Cryptographic Component Identification

The cryptographic component identification scheme presented in [HPWW05] allows to cryptographically bind all crucial vehicular components to the respective vehicle. For detection of counterfeit or bogus vehicular components, it employs small cryptographic computing tags securely (i.e., non-removable) integrated into each identifiable component and the central security module (cf. Section 7.2). The scheme further assumes that all cryptographic tags can communicate (wirelessly) with the central security module or that they can communicate (wirelessly) even directly with each other. RFID chips [HC06] could be for instance a possible realization of such wirelessly communicating cryptographic tags. The cryptographic component identification scheme, however, even works without the need of a central security module by distributing its task to all involved cryptographic tags, which are then, of course, somewhat more sophisticated and more powerful.

The life cycle of a cryptographically identifiable vehicular component is depicted in Table 8.1. The main idea is to have an initial component, for instance the central security module, that is imprinted with a secret vehicle key. Each identifiable component further holds a certificate, which is checked by the central security module at the time when a component becomes newly installed into a vehicle. On a successful certificate verification, the component becomes part of the vehicle's actual setup and receives the secret vehicle key to signal its verified legitimacy. To ensure that all identified vehicular components hold this vehicle key, in turn, a check during vehicle's operation is performed by the central security module based for instance on a challenge-response protocol (cf. Section 3.9.4). This can be done, for example, each time the vehicle is started in order to prevent manipulations even after the installation of the component. Lastly, on the removal of an identifiable component, in order to distinguish an authorized removal from a theft, the component clears the received vehicle key and, at the same time, is removed from the vehicle's actual inspection setup.

Phase	**Component installation**		**Vehicle operation**	**Component removal**
Procedure	Verification of the certificate	Establishment of the vehicle key	Check for the vehicle key	Deletion of the vehicle key

Table 8.1: Life cycle of a cryptographically identified vehicular component.

Moreover, it is possible to distribute the role of the central security module to all identifiable vehicular components. Then, all components conjointly accomplish the verification of newly installed components, check each other during vehicle's operation, and accomplish a legitimate removal. Omitting the central security module, hence, would also omit a central point of failure, but, as already mentioned before, would then require considerably more sophisticated and powerful cryptographic tags. The described cryptographic component identification scheme, moreover, is applicable not only in the automotive context, but also to many other IT systems, which consist of several components that need protection against cloning, manipulation, or theft.

8.2.2 Physically Unclonable Functions

Another promising approach for reliable vehicular component identification is based on physically unclonable functions [BCG$^+$07, GCvDD02]. Physically unclonable functions (PUF) means special combinatoric circuits integrated in a physical device (e.g., the vehicular component) that are able to provide an unpredictable random-like response by feeding the circuit with a (known) input challenge. Due to smallest individual physical differences of every physical device, even identical circuits based on identical layouts provide a very individual and unpredictable response for the same challenge. The physical identity of a PUF is based on different sources of physical randomness as result of random manufacturing variations, which are either explicitly introduced (e.g., optical or coating PUFs) or which are already inherent (e.g., delay or SRAM PUFs). Since the manufacturing process of an integrated circuit (IC) is subject to various uncontrollable and uncomputable random parameters, it is practically infeasible to create a physically identical IC (i.e., a clone) or to create a complete model of an IC to (pre-)compute the correct response. PUFs are unpredictable even for an attacker with physical access to the respective IC and are moreover resistant to invasive attacks (cf. Section 5.2.2), since even least physical modifications could affect the individual behavior of a PUF and thus its respective response.

Hence, PUFs are particularly suitable to derive a unique, unpredictable, and unclonable secret key used for dependable vehicular component identification. As described in [TB06], physically unclonable functions can be used for instance to make RFID tags and hence RFID-enabled components unclonable. However, such an approach requires effective measures that physical attacks cannot discover the (secret) response of a PUF.

8.3 Secure User Authentication

For vehicular access control schemes based on several roles or even based on particular individuals, providing a secure user authentication mechanism, which verifies a claimed user identity, is mandatory. Therefore, a user first presents his identity (or his identifier) that identifies him as a certain group member or a certain individual (*identification*). Then, the user provides a proof of truth for his claimed identity by presenting or generating the authentication information that is linked with the corresponding identifier (*verification*). An identifier for an identity could be a user name, a role/group identifier, or a certain pseudonym, which is either explicitly given or inherently assumed. Authentication information, in turn, can be classified into three different categories, namely:

- Something a person *knows* (e.g., a password)

- Something a person *possesses* (e.g., a token)

- Something a person *uniquely is* or *uniquely does* (e.g., a fingerprint or a vocal tone)

Using a secret knowledge as authentication information is simple, flexible, and cheap. However, passwords, for instance, can easily be forgotten or given away. In contrast, using a security token (i.e., something a person possesses) is more complex, more costly, and less flexible than a password. A security token further can be handed over, lost, stolen, and sometimes even copied. Biometric characteristics (i.e., something a person is or does), lastly, can be very costly and complex. Biometrics can be misinterpreted (i.e., suffer from false acceptances or false rejections) and can never be changed. However, biometric characteristics virtually cannot be forgotten, lost, stolen, copied, or given away. Lastly, for strong user authentication, information from at least two different categories (i.e., knowledge, possession, or being) are combined (e.g., presenting a smartcard together with a PIN) to provide a more reliable 2-way or a 3-way authentication, respectively.

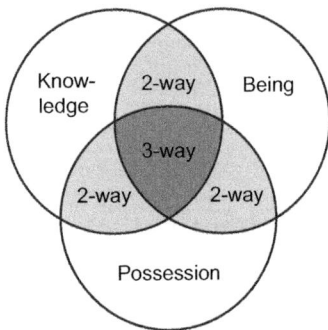

Figure 8.2: Multi-way (or multi-factor) authentication schemes based on the three general authentication information classes, namely knowledge, possession, and being.

Traditional access restricted vehicular resources, which demand for a proper authentication of the respective vehicle driver or owner are vehicle access in general, the vehicle starting, and access to certain infotainment equipment (e.g., access to the car stereo). Here, for user identification, the identifier of the driver or the owner of the vehicle is often inherently assumed without having to present a certain identity explicitly. Since solely knowledge-based authentication schemes are regarded as too cumbersome and insecure, vehicular driver authentication usually is based on security tokens such as (remote) keys or smartcards (cf. Section 4.2.2), whereas the car stereo or navigation equipment usually simply employs a personal identification number (PIN) for user authentication. Considering the fact that today most drivers already carry several very personal wireless IT devices such as mobile phones, it would be highly desirable to replace existing proprietary and incompatible mechanical key tokens with a secure mobile user authentication application, which in turn acts as a wireless vehicle key (e.g., via Bluetooth or WLAN). To prevent that the loss of the cellular results also in the loss of the vehicle keys, an additional PIN or password phrase can enable a strong and simple 2-way authentication scheme that combines possession and knowledge. Last but not least, due to the possibly unavoidable complications with lost and stolen keys (or cellulars), there exist also benefits to integrate even biometrics for driver authentication. Since at least for vehicle theft protection upper limits for corresponding costs exist, which are already met by today's vehicle keys (otherwise the vehicle will be simply picked up), biometrics can clearly enhance driver's comfort as they cannot be forgotten, lost, or stolen. Hence, for instance, fingerprint sensors could successfully emerge into the automotive domain to enhance security and comfort of the driver authentication. Moreover, a successful strong authentication scheme based on strong biometrics can be reused in other scenarios such as vehicle access to restricted areas or road pricing.

Other important vehicular areas with need for secure user authentication are vehicular diagnosis interfaces, software updates, (legally) protected vehicular settings (e.g., exhaust control), and many legal applications (e.g., milage counter) as well as various safety-critical and privacy-critical applications (e.g., event data recorder). Vehicular diagnosis devices and software flashing devices usually employ challenge-response protocols or digital certificates to verify their identity and hence to verify their authorization to access restricted vehicular functionality (cf. Section 8.4.4). Thus, the ECU authenticates the external device to be a genuine authentic OEM device, and very seldom also vica versa [MVH$^+$06, SRM06]. Vehicular applications that can have even more personal, economic, or legal impacts, such as digital tachographs or event data recorders, usually employ authentication mechanisms based on especially protected smartcards that—according to the authentication requirements of the verifying application—can implement different authentication schemes in parallel [FL06].

Lastly, secure vehicular user authentication indirectly also includes the secure transfer of authentication information to the actual verifier to prevent for instance wiretapping or modifications and hence to prevent various impersonation, replay or relay attacks (cf. Section 5.2.1). This can be realized securely by deploying for instance well-established cryptographic protocols such as challenge-response or zero-knowledge protocols (cf. Section 3.9.4).

8.4 Software Protection

As already introduced in Section 1.1 and Section 4.4, embedded vehicular software is becoming the most innovative and valuable part of current and future automotive vehicles. A today's premium vehicle has, for instance, up to several 100 megabytes of embedded code providing more than 2000 individual functions [Bro06] distributed over up to 80 individual ECUs. Future vehicles will employ more than one gigabyte of embedded code, which then will control nearly all functionality of a vehicle including such crucial functionalities such as steering or braking [CCA02, Fri04].

Ensuring highly dependable and trustworthy vehicular IT systems in the context of such a huge amount of complex software, requires adequate IT safety *and* IT security measures. As software safety measures (e.g., high redundancy, fall back mechanisms, error detection, self testing) refer to protection measures against random technical failures (*fault-tolerance*). This section provides an overview about software security measures, that means protection measures against most malicious encroachments (cf. Figure 1.2). Nevertheless, IT safety and IT security are interleaved fields, that is, vehicular software security measures can help to improve vehicular software safety and vice versa. Otherwise, it also should be ensured that a software safety measure cannot be used to realize some malicious threat and that a security measures does not become a critical safety issue.

Nonetheless, vehicular software security comprises a number of different aspects. It starts already with the software development, requires a secure software initialization as well as a secure runtime architecture, once it has been bootstrapped in the respective component. Last but not least, the software update process has to be protected against various malicious manipulations. Thus in the following, a short overview is given about the various security procedures and mechanisms to ensure the security of vehicular software.

8.4.1 Secure Software Development

Secure software development generally refers to the domains of secure software design, secure implementation, and security testing, but is also an inherent part of the organizational security (cf. Chapter 9). Hence, in the following an introduction into the most important aspects of secure software development is given. Nevertheless, further readings on this subject are helpful and can be found amongst others in [GvW03, McG06, VM02].

Secure Software Design

Security vulnerabilities resulting from architectural or design flaws are the most difficult to fix (if afterwards fixable at all) and hence the most dangerous vulnerabilities. A secure design has to prevent high-level architectural attacks such as *man-in-the-middle attacks*, *relay* and *replay attacks*, as well as various *race conditions* and *denial-of-service attacks* (cf. Section 5.2.1). Secure software design is furthermore a science on its own. However, the general steps for a secure software design process are as follows.

(1) Identify the security objectives of all entities involved (cf. Section 6.1).

(2) Deduce appropriate security requirements (cf. Section 6.2).

(3) Choose an appropriate software development process (e.g., the *Unified Process* [JBR99]).

(4) Apply a proper configuration management, that means, identify, log, monitor, and audit all activities that change the software itself, its documentation, and its tests from concept to disposal (cf. *Guidelines for configuration management* [ISO03]).

(5) Follow vulnerability discussions and corresponding publications.

Lastly, the most important principles when actually designing secure software based on [Gal98, GvW03] are the following.

- Do not make the software as secure as possible, but "just secure enough".

- Keep the design as simple as possible.

- Give an application only the *least privileges* to accomplish its objectives.

- Reuse previously tested designs.

- Choose safe default actions and default values.

- Built in an appropriate level of fault-tolerance to fail safely.

■ Do not rely on obfuscation techniques[2].

Secure Software Implementation

Security vulnerabilities resulting from implementational weaknesses are currently probably the most common security vulnerabilities [Bug08]. The most important implementational security weaknesses, in turn, are buffer overflows, stack overflows, and parsing attacks which can occur in all types of software. Figure 8.3, for instance, shows source code that employs a buffer overflow to execute an arbitrary malicious code. The exemplary code works as follows. In line 8/9 the large_string buffer is completely filled with the 32-bit address of the attacked buffer buffer from line 5. Then, in line 11/12 the arbitrary malicious code from line 1 is written at the beginning of large_string. Finally, the 128 Byte large_string is copied into the 96 Byte buffer using the unprotected strcpy function while causing a buffer overflow. This means, the original return address is overwritten with the address of the buffer itself which now has (malicious) code at its beginning.

```
1    char malcode [] = "\xeb\x1f...";
2    char large_string [128];
3
4    int main() {
5      char buffer [96];
6      long *large_string_ptr = (long *) large_string;
7
8      for ( int i = 0; i < 32; i++ )
9        *(large_string_ptr + i) = (int) buffer;
10
11     for ( int i = 0; i < strlen(malcode); i++ )
12       large_string [i] = malcode [i];
13
14     strcpy (buffer, large_string);
15   }
```

Figure 8.3: A source code example that employs a buffer overflow.

---

[2]Obfuscation is the process of trying to make reverse engineering of software (e.g., to extract valuable information concerning algorithms, data structures, or secret keys) as difficult as possible by a semantic-preserving scrambling symbols, code, and data. However, obfuscation techniques cannot prevent a really determined hacker that has plenty of time.

Again, secure software implementations are extensively discussed in litera-
ture [GvW03, HL03, McG06, VM02]. Hence, in the following just the most important issues one has to take into account when securely implementing (vehicular)
software are presented.

■ Use modern high-level languages (e.g., C++ instead of C, if possible) that
  include more security enhancing capabilities (e.g., type-safe casting, automatic array bound checking) or avoid at least insecure functions (i.e., use
  strncpy instead of strcpy or snprintf instead of sprintf).

■ Apply modern coding style guides and coding standards.

■ Keep the code as simple as possible.

■ Document code as good as possible.

■ Reuse previously tested code.

■ Validate any (external and internal) input before usage.

■ Choose safe initial values for all data.

■ Remove any obsolete code.

■ Built in proper error handling and safe degradation.

Software Security Assurance Evaluation

Software security assurance evaluations, lastly, check designs and implementations for having no vulnerable bugs, flaws, or defects. The examinations have
to be done at every development stage, best according to known methodolo-
gies [ISO05a, ISO05b, ISO07b] using a properly devised test plan, which also
includes appropriate misuse scenarios formalized in penetration tests using "abuse
cases" and "anti-requirements". Security assurance evaluations are usually done
by internal peer reviews. However, for very security-critical software they should
also be done by external (certifying) security evaluation institutions.

Besides manual code reviews, automated checking tools are particularly helpful to find security vulnerabilities in software. There exist static code checking
tools [CW07] such as Flawfinder [Whe08], Splint [E+08], or RATS [For08] that
scrutinize static software source code for potential weaknesses as well as dynamic
code checking tools such as Purify [IBM08], Valgrind [S+08], and libSafe [Ava08]
that, in turn, scrutinize corresponding binaries during runtime. Dynamic code

checking also includes profiling (e.g., using GNU's gprof [Fre08]) that enables for instance to check the test code coverage of a software, that means checking the degree to which the source code of a software is covered by existing tests.

8.4.2 Secure Software Initialization

The security of a vehicular software relies amongst others also on its correct initialization after a (temporary) deactivation, after a reset, or during the initial installation of the respective vehicular component in an unprotected environment (e.g., not within the production environment of the corresponding OEM). During the initialization (i.e., booting or loading) of a software from an unprotected location (e.g., from an external ROM or flash memory), at least its integrity has to be verified such that unauthorized (offline) modifications are infeasible or at least detectable. Further, optional security objectives could be authenticity, non-repudiation, and freshness (cf. Section 6.1). However, any secure initialization procedure therefore requires in turn a verification functionality, which is assumed to be reasonable tamper-protected (cf. Section 7.1) and is always executed first on every booting or loading process.

In the following, several different methods for secure software initialization are described together with their individual capabilities, advantages, and constraints, which, in turn, are briefly summarized in Table 8.2.

	Checksum	Hash function	MAC	Digital signature	Physical protection
Integrity	☐	■	■	■	■
Authenticity	☐	☐	■	■	■
Non-repudiation	☐	☐	☐	■	■
Freshness	☐	☐	☐	☐	■
Characteristical advantages	Simple, error-correction	Simple, fast	Simple, fast	Adequate security	Maximum security
Characteristical constraints	No security	Protected reference	Shared secret	Quite slow & complex	Most costly & inflexible

Table 8.2: Overview of several methods for secure software initialization together with their individual capabilities, advantages, and constraints.

Checksums

The term checksum usually means additional information (e.g., parity bits, or check digits) added to data for detection (and correction) of accidental alterations such as data corruptions. However, even sophisticated checksums such as cyclic redundancy checks [PB61], cannot provide any security against malicious encroachments, since an attacker can easily recompute and adapt checksums due to their trivial mathematical structure and the absence of any inherent secrets. Thus, checksums can improve safety by detecting random technical failures, but cannot be used to secure software initializations against malicious encroachments.

Cryptographic Hash Functions

Cryptographic hash functions such as MD5, SHA-1, or RIPEMD (cf. Section 3.5) are keyless one-way functions, which are able to ensure information integrity, that means, which are able to detect errors and malicious manipulations. For integrity verification, hash functions compress data of any length to a small (e.g., 160 bit) unique string of fixed length, the so-called hash value or digital fingerprint, which is then compared with a (protected) reference value. Due to their one-way characteristic, it is computationally infeasible to generate a different information (or software) where the hashing process produces the same hash value. Hence, for securing software initializations, hash functions can reliably ensure software integrity. They furthermore can be implemented to run fast and efficiently in hardware and in software. However, since a hash function does not include any secrets (e.g., secret keys), an attacker can always compute a valid (but different!) hash value also for a manipulated software. Thus, for a reliable software integrity enforcement, the secure initialization function requires secure access to the corresponding protected reference values.

Message Authentication Codes

As described in detail in Section 3.6, Message authentication codes (MACs) basically are cryptographic checksums based on adapted symmetric block ciphers, for example as defined in [ANS95a] that employs an adapted TripleDES, or are based on keyed hash functions (HMAC). Similar to cryptographic hash functions, MAC functions compress data of any length to a small unique string of a fixed length, the MAC. However, in contrast to keyless hash functions, MAC functions additionally integrate a shared secret key, so that only holders of this secret key are able to create and verify the respective MAC. Thus, additionally to the integrity verification of information, MACs also enable the authentication of the information origin, which means they can verify that the creator of a certain MAC has been in the possession of the shared secret key. Hence, for securing vehicular software initializations, MAC functions can reliable ensure software integrity and software authenticity. They can be implemented fast and efficiently in hardware and in software. However, all parties involved in a MAC-based software integrity protection scheme have to hold and have to protect the shared secret key individually. MAC schemes further cannot provide non-repudiation, since every entity involved can create indistinguishable MACs using the shared secret key.

Digital Signatures

As shown in Section 3.9.1, a digital signature is an asymmetric cryptographic scheme based on a hash function and asymmetric primitives. The integrated hash function compresses the data to a unique hash value, which is then signed with an asymmetric algorithm using a secret private key. The encryption result, that means the digital signature, is then appended to the respective data. For signature verification, the verifier applies the corresponding public key on the digital signature and checks if the result matches with the independently computed hash value of the data. Thus, only the holder of the secret key is able to sign an information, whereas anyone can verify the signature of an information by using the corresponding public key. Digital signatures are able to ensure integrity and authenticity of a software, but, in contrast to MACs, do not rely on a shared secret. They further are able ensure non-repudiation, since only the holder of the secret key is able to sign a software. However, digital signature implementations are comparatively costly and slow in hardware as well as in software. At least some adapted verification schemes, for instance based on RSA [Hås85], can be implemented quite efficiently. Digital signatures further do not inherently provide any freshness detection. But nevertheless, having the public verification key nonexchangeably linked with the initialization function (e.g., as part of the protected verification functionality), digital signatures can be used to reliably ensure secure initialization of vehicular software components.

Physical Protection

By employing a completely hardware-protected software initialization that enforces integrity, authenticity, non-repudiation, and even freshness [3] without having

---

[3]Freshness means any kinds of replay attacks are infeasible such that the software being loaded is always the one that has been written at last.

to rely on cryptographic measures, the highest level of security can be achieved. Further, hardware-based software protection mechanisms are usually quite fast and easy to use, since they do not have to deal with, for example, shared secrets, certificates, or public key infrastructures.

However, employing tamper-protected hardware (cf. Section 7.1) is normally the most costly approach and, due to its immutable nature, often also quite inflexible. Nevertheless, in the end, every cryptographic mechanism usually also relies on a small piece of trusted hardware, which physically protects the essential security-critical operations and corresponding secrets. Hence, initializing for instance a small immutable piece of software from a physically protected read-only memory, which then in turn protects the further software initialization from unprotected storages, can be both adequately efficient and sufficiently secure.

Verifiable Initialization (TCG)

Lastly, a short description of the Trusted Computing based approach for verifiable software initialization is given, which combines cryptographic hash functions and public key cryptography within a protected TPM chip to establish a *verifiable chain of measurements* [Tru07b]. As introduced in Section 3.8.1, in this approach a TPM-enabled BIOS represents the physically protected *Core Root of Trust for Measurement*, which securely *measures*, that means computing its cryptographic hash value, the software component executed first (e.g., the master boot record) before passing control to it. The firstly measured and executed software component then again employs the TPM to measure the software component executed next (e.g., the boot loader). Thus, a verifiable chain of measurements is established such that, before execution, each software component is measured by a previously measured and executed software component.

However, the measurement (i.e., the computation of the hash values), the storage, as well as the provision of the respective hash chain are securely managed by the TPM and hence protected against any software attacks. Upon completion of a verified software initialization process, these measurement values reflect the *configuration* of the currently running hardware and software environment. Trusted Computing technology, however, remains passive and does explicitly not prevent a certain computing environment from being compromised during runtime. Nevertheless, it makes possible manipulations detectable and allows a TPM-enabled operating system to react according to its effective security policies. Further TCG capabilities such sealing, binding, or attestation (cf. Section 3.8.2) allow to link the access to secrets or keys protected by the TPM chip to certain configurations or to verifiable report the actual configuration to local and external parties.

8.4.3 Software Security Architectures

A software security architecture enables a single computing devices to securely run multiple (independent) processes in parallel on top of a system structure that enables sharing of available resources. The most important safety and security requirement such a software security architecture has to provide is *strong isolation* (also called *runtime isolation*) that ensures that subsystems, components, or even individual applications can communicate only via strictly controlled communication channels such that it is impossible to access (i.e., data, functionality) or even affect (e.g., performance) each other without proper authorization. A software security architecture further provides some essential security mechanisms such as secure storage or secure communication according to the individual application-specific security requirements.

To efficiently regard the individual application-specific security requirements and the specific technical constraints of a critical vehicular application, different software security architectures are possible. The traditional approach are *monolithic security architectures*, which make verifications for correctness complex and costly, and hence, are likely to include flaws and vulnerabilities. *Hardware isolated security architectures*, however, require sophisticated security modules that provide a hardware-based separation/virtualization functionality. *Virtualized security architectures*, lastly, ensure strong isolation by employing a very small, and hence easy to verify, hardware-based and/or software-based separation/virtualization mechanism. The individual characteristics of the mentioned software security architectures are introduced in the following subsections.

Virtualized Security Architecture

Figure 8.4 provides a general approach to a flexible multi-purpose security architecture that is loosely based on the PERSEUS security architecture [PRS$^+$01], where a small security kernel serves as a control instance between the vehicular hardware and the actual application layer. Here, a small security kernel provides strong separation of resources and implements elementary security mechanisms. According to the ongoing development from today up to 80 individual functionally independent ECUs [Bro06] to future vehicles that will employ only three or five powerful multi-purpose controllers [CCA02, Fri04], this security architecture is reasonable and realistic. In the following, all four layers of the virtualized vehicular security architecture depicted in Figure 8.4 are described briefly.

Hardware Layer. The hardware layer consists of conventional vehicular hardware extended by a security module (cf. Section 7.2) which is separately protected by

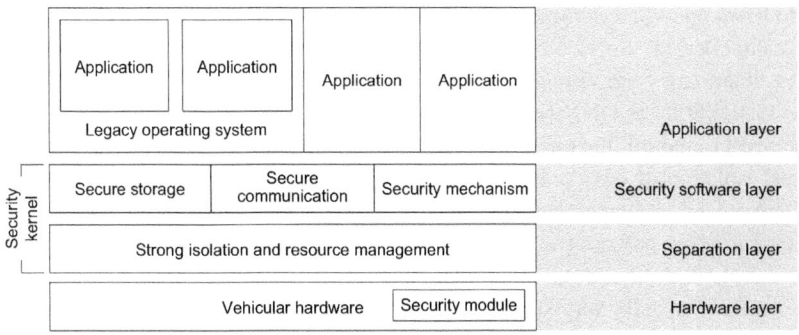

Figure 8.4: Virtualized vehicular security architecture.

physical security measures (cf. Section 7.1). The security module usually is a dedicated cost-effective cryptographic microchip securely attached to the vehicular hardware that provides fundamental security-critical functions such as cryptographic operations (e.g., encryption, decryption, or hashing), secure timing or secure random number generation. Hence, security-critical information (e.g., secret keys) are securely protected by hardware measures and cannot be affected by any (malicious) software functionality. During startup of such a vehicular multipurpose controller, the hardware layer could enable the verifiable or secure initialization by cryptographically measuring (e.g., by a cryptographic hash function) or cryptographically verifying (e.g., by a digital signature) the corresponding bootstrapping process.

Separation Layer. The separation management layer provides an abstraction of the underlying hardware (e.g., microprocessors, interrupts, devices) and implements essential OS functionalities such as threads, logical address spaces, and inter-process communication (IPC). It further provides an appropriate resource management interface and enforces the effective access control policy based on these resources. The separation management layer particularly ensures strong isolation such that applications, components, or subsystems can be isolated from each other, that is, they can communicate with each other only via strictly controlled communication channels. The separation management layer can be realized by a conventional monolithic vehicular operating system such as OSEK-OS [OSE08], RTLinux [YB97], or VxWorks [Win08]. However, similar to monolithic security applications, because of the inherently large complexity, monolithic operating systems are hardly fully verifiable for correctness and hence are likely to include flaws and vulnerabilities in design or implementation that can be exploited by an attacker (cf. Section 5.2.1). Another approach for realizing the separation management layer are virtualization technologies based on microkernels [4] such as PikeOS [SYS08] or QNX [Hil92] or based on a hypervisor [5] as Xen [BDF+03]. In contrast to monolithic operating systems, virtualization technologies enable the mutual isolation of device drivers and other essential operating system services, such as process management and memory management, from the actual security policy enforcing process (i.e., the microkernel or the virtual machine monitor respectively). Thus, the actual amount of code which has to be securely verified for correctness is usually very small, and a failure in one of the isolated OS services cannot directly affect the other services, especially not the security policy enforcing process. Virtualization technology moreover enables reutilization of legacy operating systems and existing applications.

Security Software Layer. The security software layer uses the functionality offered by the separation layer to provide security functionalities on a more abstract level. It provides elementary security mechanisms such as secure storage or secure communication. The security mechanisms are realized independent of and protected from the respective applications, which again reduces verification complexity and hence improves reliability. Some essential security mechanisms of the security software layer (i.e., secure storage, secure communication) are described in more detail in the subsequent sections.

Application Layer. On top of the security kernel, several (security-critical) applications and, if virtualization technology is available, multiple legacy operating systems can be executed concurrently, but strongly isolated. Communication between applications as well as potential I/O access are subject to the conditions of the effective security policy enforced by the underlying security kernel. The application layer provides all services and (parts of) applications that are not security-critical as well as the corresponding user interfaces.

Monolithic Security Architecture

However, for dedicated isolated security applications, which employ the available hardware exclusively, it is nevertheless possible to even use a minimized security kernel well adapted only to the (security) requirements of the actual applications.

---

[4] A microkernel is an OS kernel that minimizes the amount of code running in privileged processor mode, since device drivers and other essential operating system services run in isolated user-mode processes.

[5] A hypervisor or a virtual machine monitor (VMM) allows multiple operating systems (virtual machines) to run concurrently and independently on a single host platform.

As depicted in Figure 8.5, the security kernel provides only basic resource management services and the security mechanisms that are essential for the respective (security) application without any further mechanisms for resource separation. Monolithic applications that themselves include all necessary resource management and security mechanisms could even completely omit a separate security kernel, while running directly on top of the dedicated hardware. However, due to the inherently large complexity of such monolithic applications, they are hardly completely verifiable for correctness and hence are likely to include flaws and vulnerabilities in design or implementation that can be exploited by an attacker (cf. Section 5.2.1).

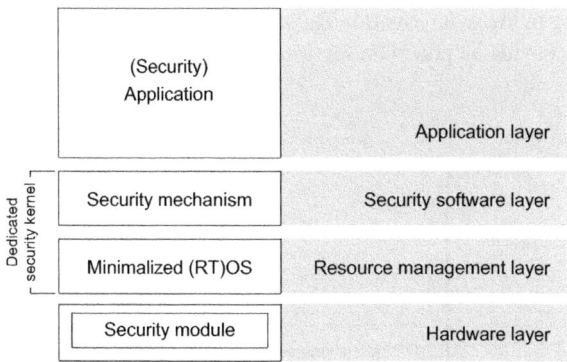

Figure 8.5: Dedicated monolithic vehicular security architecture.

Hardware Isolation

Hardware isolated software security architectures rely on particular hardware isolation mechanisms such as ARM TrustZone Technology (TZ) [AF04, WFM+07] or Intel's Trusted Execution Technology (TXT) [Int07]. Instead of running a dedicated security kernel that handles all separations in software, these approaches have some small additional hardware mechanisms built-in, which then virtually provide an additional secure execution environment, which is strictly isolated from the legacy execution environment by dedicated hardware measures. Thus, the processor can switch from the legacy environment (*non-secure world*) to a virtual security environment (*secure world*) such that each world operates independently, but in parallel on the same processor. Coupled hardware access control measures for memory access and corresponding peripherals thereby prevent that information can leak from the secure world into the non-secure world. Communication between the two worlds is possible only by calling a privileged instruction from the non-secure legacy kernel. When employing ARM's TrustZone technology for instance, calling the privileged *secure monitor call* (SMC) instruction switches the processor from the non-secure world to the secure world and passes all enclosed information to a protected security monitor. The *secure monitor*, in turn, has tightly controlled entry points and makes no assumptions on possible data structures to prevent all kinds of buffer overflows. To switch back to the non-secure world, the security kernel sets a specific status bit, which also determines whether a program executes in the secure or non-secure world. Devices attached to the core can also be marked as secure or non-secure such that secure devices can only be controlled by a driver running in the secure world. Thus, hardware isolated software security architectures can provide an adequate security level for security-critical vehicular applications.

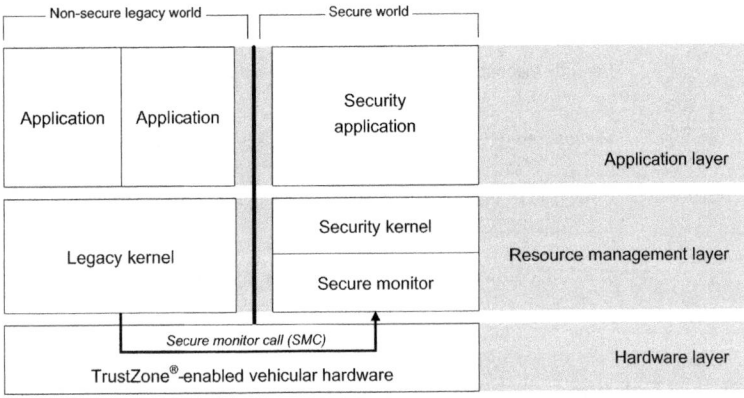

Figure 8.6: Hardware isolated vehicular security architecture exemplified by the ARM TrustZone Technology.

However, as already mentioned in the section's introduction, hardware isolation measures require somewhat sophisticated (and thus more costly) security modules and, in contrast to virtualized security architectures, provide only limited isolation between security processes within the secure world (and also between processes within the non-secure world). Thus, a failure in one of the security applications can directly affect the other security applications.

8.4.4 Secure Software Updates

As described in detail in Section 4.4, embedded software is becoming the most innovative and valuable part of current and future automotive vehicles. Due to the ubiquitous use of flashable ECUs[6], software updates can take place also after delivery of the vehicle. However, since vehicular software usually implies various copyright, safety, liability, warranty, or business issues, software updates are particularly susceptible to malicious encroachments and hence particularly require proper protection measures. As described in detail in Section 8.4.2 about secure software initialization, simple safety-minded approaches usually cannot meet the security objectives for vehicular software updates (cf. Section 4.4). Thus, in the following, two mechanisms for securing software updates (or after-sale feature activations) are presented. First, a general approach based on digital signatures is described, followed by a more extended approach based on Trusted Computing technology.

General Approach based on Digital Signatures

As shown in Section 3.9.1, digital signatures based on asymmetric cryptography are a powerful method to enforce authenticity, integrity, and non-repudiation of digital information. In short, digital signatures are based on an asymmetric key pair consisting of a secret private key for signature generation and a public key for signature verification[7]. Thus, only the holder of the secret key is able to sign an information, whereas anyone can verify the signature by using the corresponding public key. Digital signatures are already used in many applications such as electronic mails or electronic banking to provide information integrity and information authenticity. Thus, digital signatures are also a powerful technology to protect software updates in vehicles [MVH+06].

Figure 8.7 exemplarily shows the steps involved for securing software updates in the automotive domain protected by means of digital signatures. After the software development (1), the software is signed at the OEM's trust center (2). The trust center of the OEM is assumed to be a protected environment (cf. Chapter 9), which securely employs the OEM's secret key to authenticate authorized software by creating the corresponding digital signature. The signature is appended to the software (3) and stored together in the OEM's database for further distribution and

---

[6]A flashable ECU is a microcontroller capable of reprogramming its memory for application programs and data based on so-called flash memory technology [MVH+06].

[7]In contrast to asymmetric encryption, where the sender encrypts a message with the public key of the receiver such that it can be decrypted again only by the receiver, who has the corresponding private key.

provision for flashing (4). The signed software then can be loaded into the respective flashing tool (5), while the flashing tool itself can be authenticated optionally through the database by checking the digital certificate[8] of the flashing tool (if available). During the setup for a (freshness preserving) secure channel[9] between the flashing tool and the respective ECU (7), the flashing tool again can prove its authenticity against the ECU by providing an appropriate digital (OEM) certificate. Lastly, the ECU can verify the loaded software for integrity and authenticity by verifying the corresponding signature using the corresponding public key securely protected inside of a security module (6). After successful verification, the ECU can safely install the respective software update.

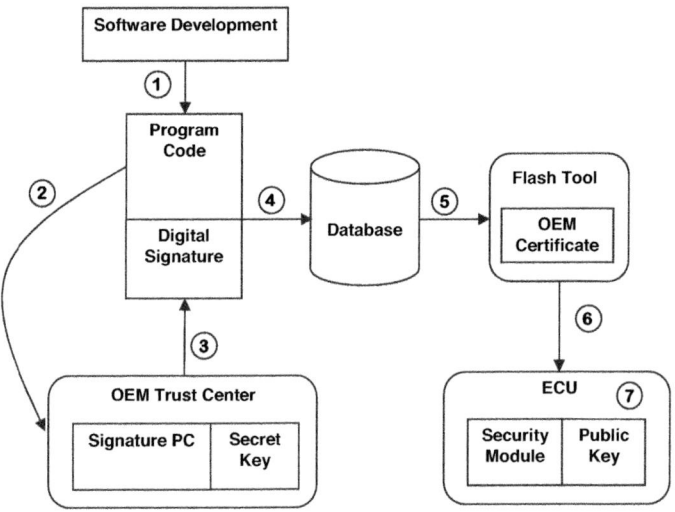

Figure 8.7: Steps involved for securing software updates in the automotive domain protected by means of digital signatures.

The overhead of this approach is rather low. A typical algorithm for embedded signature verification (e.g., RSA [RSA78]) requires around 5 kB of ECU's memory. Besides the program memory, the ECU has to store the respective

---

[8]A digital certificate means an electronic file that again incorporates a digital signature from a trusted authority (e.g., the OEM) that verifies the authenticity of its holder.

[9]A secure channel ensures confidentiality and integrity of the communicated data as well as the authenticity of communication endpoints. Note, establishing a secure channel between the flashing tool and the ECU involves additional cryptographic measures, which are addressed in Section 3.9.2.

public key using integrity-preserving read-only memory, which cannot be modified or exchanged by an potential attacker. However, the described approach does not provide confidentiality of the software update, which requires additional cryptographic measures based for instance on symmetric cryptography schemes (cf. Section 3.2). Hybrid encryption schemes (cf. Section 3.9.2), for example, ensure confidentiality and can easily be integrated into the afore described digital signature based approach. Nevertheless, further reading on secure software updates in embedded systems is helpful and can be found amongst others in [AHS05, MVH$^+$06, SRM06, ZNK$^+$06].

Trusted Computing based Approach

The extended approach for securing vehicular software updates is based on virtualization technology and Trusted Computing (cf. Section 3.8). The proposed scheme allows a vehicular content provider to "bind" arbitrary software and digital content to a certain vehicle hardware and software configuration. In other words, a content provider can reliably ensure that his bound (and thus encrypted) content can be decrypted and accessed only by a previously authorized vehicle hardware and software configuration while being distributed using a fully untrusted infrastructure. This binding functionality (cf. Section 3.8.2) can be used to limit the application of a software update or the usage of a digital content to a certain vehicle brand, a certain vehicle type, or even to an individual car. Moreover, since decryption of bound content is possible only on previously defined hardware and software configurations, a content provider can enforce that its content is used only on platform configurations that implement appropriate know-how theft prevention and/or Digital Rights Management (DRM) capabilities so that decrypted content cannot leak into untrusted software.

The described approach employs a TPM-based security module as described in Section 7.2.3 and a virtualized security architecture as described in Section 8.4.3, that means, all security software components are running in parallel to, but strongly isolated from, other legacy applications or other operating systems on top of a small security kernel. The two (abstract) parties involved in this protocol are the *vehicle user* and the *content provider*. The content provider distributes digital content (ECU software, navigation data, media files, etc.) that will be applied by the vehicle user. Thus, the role of the content provider exemplarily summarizes original equipment manufacturers, authorized suppliers, infotainment data vendors and similar parties that are authorized to provide vehicular content. Whereas, the role of the vehicle user refers to the person that currently uses the vehicle or simply to the actual vehicular IT system.

As simplified illustrated in Figure 8.8, the protocol basically consists of four steps. First, the TPM securely retrieves the current vehicle hardware and software configuration (cf. Section 3.8.2) and creates an asymmetric key pair together with a certificate, stating that the corresponding private key is securely stored inside the TPM and its usage is bound to the current vehicle configuration (1). Secondly, the certificate and the public key are transferred to the content provider, who validates the certificate and hence validates the corresponding vehicle hardware and software configuration. This validation implies verifying the digital signature and checking the vehicle configuration against "known good" reference values (2). Thirdly, on successful validation, the content provider uses the public key included in the certificate to establish a *trusted* channel (cf. Section 3.8.2) to the vehicle (3). This means that any content encrypted using this public key can be decrypted only if the vehicle provides the *trusted* configuration as stated in the afore successfully validated certificate. A trusted configuration in this context means that the virtualized security architecture is running, and that all components (both hardware and software) processing protected content are trusted by the respective content provider. The bound and decrypted content is finally sent to the vehicle for decryption and usage if (and only if) it provides the stated trusted configuration (4). Note that steps (1) and (2) can be done already at the end of the manufacturing process by the OEM, thus reducing initial complexity. A new asymmetric key pair and the corresponding certificate is thereafter only necessary if one of the hardware or software components, which are part of the verified vehicle configuration, has been modified due to a software update or due to a hardware change.

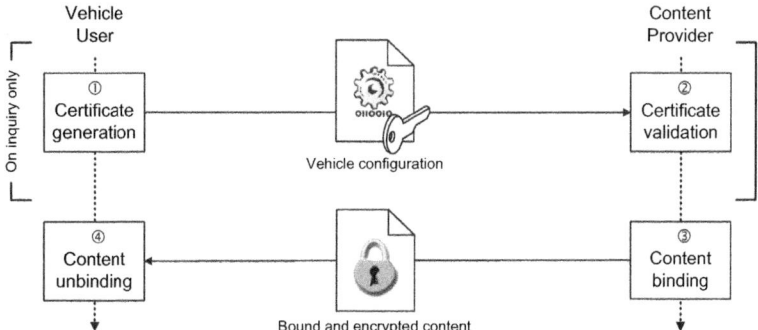

Figure 8.8: Steps involved for securing software updates in the automotive domain protected by means of Trusted Computing and a virtualized security architecture.

To employ this approach for securing software updates, the content provider side requires no changes of their existing server environment (and does not need Trusted Computing technology either). Thus, they can employ their common computing architecture to accomplish certificate validation (2) and content binding (3). As exemplarily shown in the corresponding protocol for content unbinding in Figure 8.9, the software components involved in certificate creation (1) and content unbinding (4) on vehicle user side are basically a Trusted Computing service (*Trust Manager*), which provides an abstraction of available Trusted Computing functionality, an advanced software loader (*Compartment Manager*), which manages the integrity values of all software components loaded, and the respective rendering application (*Trusted Viewer*), which decrypts the protected content and makes use of it. Both have to be strongly isolated and protected from other applications and available legacy operating systems by the means of the virtualized security architecture to prevent unauthorized access and any malicious encroachments. Further details on the protocol and the actual implementation can be found in [SSW06].

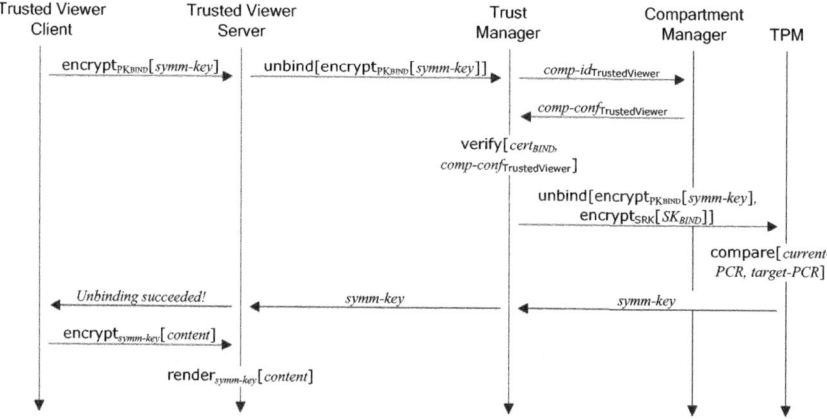

Figure 8.9: Content unbinding protocol executed at the vehicle (i.e., step (4) of the overall software update protection procedure shown in Figure 8.8) where the Trust Manager first verifies the integrity of the Trusted Viewer application before it requests the TPM to unbind the symmetric encryption key, which in turn is bound on the overall vehicular hardware and software configuration.

8.5 Secure Storage

Secure storage enables vehicular applications to securely save and to securely retrieve their critical information such as personal information (e.g., navigation routes, communication data), operational secrets (e.g., authorizations, check sums, mileages), or technical secrets (e.g., sophisticated algorithms, special setups, protected know-how) to and from a persistent storage location. Secure storage basically protects critical vehicular information against offline attacks (cf. Section 5.2), that means, while the respective vehicular component is (temporary) deactivated. It thus can prevent unauthorized access, unauthorized manipulations, unauthorized copies, privacy violations or information theft in case the respective vehicular component is demounted, stolen, or otherwise compromised while not being in operation[10]. Briefly summarized, information handled by a secure storage mechanism is subject to the following security objectives.

Confidentiality of data. Unauthorized access to protected information is infeasible.

Integrity of data. Unauthorized alterations of protected information is at least detectable.

Availability of data. Authorized entities (e.g., applications, users) have proper and timely access to their information.

Freshness of data. Any kinds of replay attacks are infeasible, that means, the information being retrieved is always the one that has been saved at last.

One approach to realize a secure storage application that is able to fulfill all these security objectives would be a a fully tamper-protected storage component with a proper authentication interface (cf. Section 8.3). However, this approach has turned out to be too costly and inflexible even in the world of general-purpose computing, and thus, would be even more inapplicable in a vehicular environment. Hence, most approaches for realizing secure storage are based on a combination of cryptographic measures, that means, a symmetric encryption algorithm (confidentiality) together with a cryptographic hash function (integrity) extended by appropriate anti-replay (freshness) and redundancy (availability) measures. In contrast to a fully tamper-protected storage component, the cryptographic approach

---

[10]Online or runtime protection of critical vehicular information cannot be provided by a secure storage mechanism and has to be ensured by appropriate operating system and/or further hardware and software security measures (cf. Section 8.4).

reduces the security demands for a reliable secure storage implementation to the following.

■ Confidentiality of the encryption/decryption secret.

■ Freshness of the freshness verification reference.

■ Protected cryptographic operations.

■ Secure authentication interface.

This means all cryptographic secrets (e.g., decryption keys) and necessary reference values (e.g., hash values, freshness counters) as well as all operations that employ these pieces of information (e.g., encryption, decryption, integrity verification, freshness verification) have to be stored and have to be implemented in a protected environment respectively. However, most vehicular components are under the full logical and physical control of their owners, who can attack and circumvent even sophisticated (software) protection mechanisms by running exploits, resetting components, mounting replay attacks, or reconfiguring underlying layers and operating systems (cf. Section 5.2). Thus, implementing a cryptography-based secure storage in a vehicular environment requires adequate hardware-based protection measures. However, the protective measures for the encryption keys, the reference values, and the corresponding cryptographic operations can be effectively provided by a proper security module (cf. Section 7.2). Finally, the secure authentication of users or applications to prevent unauthorized access to protected information is usually based on the verification of an authorization secret, but can be further extended by a Trusted Computing functionality named *sealing* (cf. Section 3.8.2). Similarly to the binding of software updates as described in Section 8.4.4, the information can additionally be "sealed" to a certain vehicle hardware and software configuration. In doing so, sealed information can be "unsealed" (i.e., decrypted and accessed) only by exactly the same vehicle hardware and software configuration as it was during the sealing process. Thus, manipulations on parts of the vehicle configuration that are verified during the unseal process render sealed information inaccessible. Further readings on a such an implementation can be found in [SSS⁺06, SSSW06].

8.6 Secure Communication

As described in detail in Section 4.7, vehicular communication (VC) can be distinguished into in-vehicle communication, that means communication between different ECUs distributed all over the vehicle, external wired communication such as vehicle-to-device (V2D) communication, as well as external wireless communication such as vehicle-to-infrastructure (V2I) communication, and vehicle-to-vehicle (V2V) communication. The security objectives of VC applications mentioned in Section 4.7 comprise both technical and organizational aspects and can basically be summarized to the following (even though, not all security objectives are necessarily relevant to all VC applications).

Message confidentiality. Unauthorized access to contents of protected messages is infeasible.

Message integrity. Unauthorized alterations of messages can be at least detected.

Message freshness. Any kinds of replay attacks are infeasible.

Message authenticity. The origin of a message must be verifiable.

Message availability. Authorized communication entities must have proper and timely access to their messages.

A further overall security related objective is *privacy* such that the usage of any vehicular communication application must not endanger the privacy policies of the respective driver/owner (cf. Section 4.9). However, privacy protection is mainly based on the overall design of a vehicular application and on general organizational measures, whereas message confidentiality, for instance, is a feasible security objective in order to realize one aspect of privacy protection in VC applications. Moreover, a few VC applications further require *non-repudiation* that enables communication nodes to prove to a third party that one node has actually generated a certain message (e.g., for some legal or liability issues).

Nevertheless, every VC application has its very individual security objectives, often even for every message. A warning message, for instance, should be authentic, but normally does not require confidentiality. In the following, an introduction of possible security mechanisms and measures that enable a VC application to fulfill the above-mentioned security objectives for vehicular communication is presented. Nevertheless, further readings are helpful and can be found amongst others in [Fib04, PWW04b, PWW04c] regarding in-vehicle vehicular communication security, in [GFL+07, HCL04, IEE06b, GST05, KMS06, PGH06, PP05, RH07, RPH06] regarding external vehicular communication security, as well as in [PW06] considering the security implications when deploying currently available wireless communication protocols such as GSM or UMTS for vehicular communication applications.

8.6.1 In-Vehicle Communication Security

As described in detail in Section 4.7.1, today vehicles already contain a multiplicity of controllers and sensors that are increasingly networked together by various internal communication systems with very different properties and capabilities. Hence, in the following, first a general overview about the properties and capabilities of current typical in-vehicle communication systems on the basis of four representative exemplars is given. Then, in contrast to the more general attack descriptions and exposures described in Section 5.2.1, some more detailed descriptions of possible in-vehicle communication attacks based on the the four examples introduced before are presented. Lastly, this section gives an introduction about possible security mechanisms and applicable security protocols to enable in-vehicle communication security.

In-Vehicle Communication Networks

In contrast to point-to-point communication, the bus-driven in-vehicle communication network logically connects several peripherals using the same set of wires. This enables various synergy effects (e.g., signal sharing, cost savings, weight savings) and inherently reduces fuel consumption and maintenance requirements. In-vehicle communication networks are further easy to implement and easy to extend, and are just mandatory to realize several complex networked vehicular applications. According to their essential technical properties and application areas, VC networks can be distinguished in five different groups (cf. Table 8.3). Local subbus networks such as LIN (Local Interconnect Network) control small autonomous networks used for automatic door locking mechanisms, power-windows and mirrors as well as for communication with miscellaneous smart sensors to detect, for instance, rain or darkness. Event-triggered bus systems like CAN (Controller Area Network) are used for soft real-time in-car communication between controllers, networking for example the anti-lock breaking system (ABS) or the engine management system. Time-triggered hard real-time capable bus systems such as FlexRay, TTCAN (Time-Triggered CAN) or TTP (Time-Triggered Protocol) guarantee determined transmission times for controller communication and therefore can be applied in highly safety relevant areas such as in most Drive-by-Wire systems. The group of multimedia bus systems like MOST, D2B (Domestic Digital Bus) and GigaStar arise from the new automotive demands for in-car entertainment that needs high-performance, wide-band communication channels to transmit high-quality audio, voice and video data streams within the vehicle. The wireless communication group contains modern wireless data transmission tech-

nologies that more and more expand also into the automotive area. They enable
the internal vehicle network to communicate with other cars nearby, external base
stations as well as the utilization of various location based services.

Subbus	Event-triggered	Time-triggered	Multimedia
LIN	CAN	FlexRay	MOST
K-Line	VAN	TTP	D2B
I^2C	PLC	TTCAN	GigaStar

Table 8.3: Grouping of selected in-vehicle bus communication systems.

In the following, a short technical description of an appropriate representative
from each identified vehicular communication network group, whereas Table 8.4
summarizes the corresponding characteristics. However, further readings can be
found amongst others in [Fle00, ISO06, LIN99, MS07, MOS98, ZS07].

LIN. The UART[11] based Local Interconnect Network [LIN99] is a single-wire sub
network for low-cost, serial communication between smart sensors and actuators
with typical data rates up to 20 kbit/s. It is intended to be used from the year 2001
on everywhere in a car, where the bandwidth and versatility of a CAN network is
not required. A single master controls the collision-free communication with up to
16 slaves, optionally including time synchronization for nodes without a stabilized
time base. LIN is (similarly to CAN) a receiver-selective bus system. Incorrect
transferred LIN messages are detected and discarded by the means of parity bits
and a checksum. Beside the normal operation mode, LIN nodes provide also a
sleep mode with lower power consumption, controlled by special sleep (respectively wake-up) message.

CAN. The all-round Controller Area Network [ISO06], developed in the early
1980s, is an event-triggered controller network for serial communication with data
rates up to one Mbit/s. Its multi-master architecture allows redundant networks,
which are able to operate even if some of their nodes are defect. CAN messages
do not have a recipient address, but are classified over their respective identifier.
Therefore, CAN controller broadcast their messages to all connected nodes and
all receiving nodes and decide independently if they process the message. CAN
uses the decentralized, reliable, priority driven CSMA/CD (Carrier Sense Multiple Access / Collision Detection) access control method to guarantee always the
transmission of the top priority message first. In order to employ CAN also in the

---

[11]Universal Asynchronous Receiver Transmitter

environment of strong electromagnetic fields, CAN offers an error mechanism that detects transfer errors, interrupts and indicates the erroneous transmissions with an error flag and initiates the retransmission of the affected message. Furthermore, it contains mechanisms for automatic fault localization including disconnection of the faulty controller.

FlexRay. FlexRay [Fle00] is a deterministic and error-tolerant high-speed bus, which meets the demands for future safety-relevant high-speed automotive networks. With its data rate of up to 10 Mbit/s (redundant single channel mode) FlexRay is targeting applications such as Drive-by-Wire and Powertrain. The flexible, expandable FlexRay network consists of up to 64 nodes connected point-to-point or over a classical bus structure. As physical transmission medium both optical fibers and copper lines are suitable. FlexRay is (similarly to CAN) a receiver-selective bus system and uses the cyclic TDMA (Time Division Multiple Access) method for data transmission control. Therefore, it uses synchronous transmission for time-critical data and priority-driven asynchronous transmission for non-time-critical data via freely configurable, static and dynamic time segments. Its error tolerance is achieved by channel redundancy, a protocol checksum and an independent instance (bus guardian) that detects and handles logical errors.

MOST. The ISO standardized Media Oriented System Transport [MOS98] serial high-speed bus became the basis for present and future automotive multimedia networks for transmitting audio, video, voice, and control data via fiber optic cables. The peer-to-peer network connects via plug-and-play up to 64 nodes in ring, star or bus topology. MOST offers, similarly to FlexRay, two freely configurable, static and dynamic time segments for the synchronous (up to 24 Mbit/s) and asynchronous (up to 14 Mbit/s) data transmission, as well as a small control channel. The control channel allows MOST devices to request and release one of the configurable 60 data channels. Unlike most automotive bus systems, MOST messages include always a clear sender and receiver address. Access control during synchronous and asynchronous transmission is realized via TDM (Time Division Multiplex) respectively CSMA/CA. The error management is handled by an internal MOST system service, which detects errors over parity bits, status flags and checksums and disconnects erroneous nodes if necessary.

For network spanning communication, automotive bus systems require appropriate *bridges* or *gateways* to transfer messages among each other despite their different physical and logical operating properties. Gateways read and write all the different physical interfaces and manage the protocol conversion, error protection and message verification. Depending on their application area, gateways include sending, receiving and/or translation capabilities as well as some appropri-

	LIN	**CAN**	**FlexRay**	**MOST**
Adapted for	Low-level non-real-time subnets	Soft real-time control networks	Hard real-time control networks	Multimedia, telematics
Application examples	Door locking, climate control, power windows, light/rain sensor	Antilock breaking, driving assistants, engine control, electronic gearshift	Break-by-wire, steer-by-wire, shift-by-wire, emergency systems	Entertainment, navigation, information, mobile Office
Physical layer	Single-wire	Dual-wire	Optical fiber, dual-wire	Optical fiber
Architecture	Single-master	Multi-master	Multi-master	Multi-master
Access control	Polling	CSMA/CA CSMA/CR	FTDMA, TDMA	CSMA/CA, TDM
Transfer mode	Synchronous	Asynchronous	Synchronous, asynchronous	Synchronous, asynchronous
Data rate	20 kbit/s	1 Mbit/s	10 Mbit/s	24 Mbit/s
Relative cost to CAN node	0.5	1	2.5	5
Safety measures	Checksum, parity bits	Checksum, parity bits	Redundant channels, bus guardian	Checksum, system service
Security measures	None	None	None	None

Table 8.4: Capabilities of some selected in-vehicle bus communication systems [Fle00, ISO06, LIN99, MOS98].

ate filter mechanisms. While so-called super-gateways interconnect centralized all existing bus systems, local gateways link only two different bus systems together. Therefore, super gateways require some kind of sophisticated software and plenty of computing power in order to accomplish all necessary protocol conversions, whereas local gateways realize only the hard- and software conversion between two different bus backbones.

Exemplary Exposures of In-Vehicle Communication Networks

As already mentioned in Section 5.2.1 and shown in Table 8.4, internal vehicle communication networks assure safety against several technical interferences, but they are usually unprotected against any malicious encroachments. The increasing coupling of in-vehicle networks with user-access multimedia busses or even with external wireless networks cause various additional security risks [MZ05], which have been hardly analyzed [PWW04b]. As shown in Table 4.1, the consequences of successful in-vehicle communication attacks range from minor comfort restraints up to the risk of an accident. While attacks on LIN or multimedia networks may result in the failure of power windows or navigation software, successful attacks on CAN networks may result in malfunction of some important driving assistants, that leads to serious impairments of the driving safety. A succeeded systematic malfunction on hard real-time buses like FlexRay, which handle essential driving commands like steering or breaking, can lead in acute hazards for the affected occupants and other road users. Nevertheless, even a simple malicious car locking may have serious consequences for affected occupants [Ban03].

Additionally to the general technical and organizational constraints (cf. Section 6.4.1 and Section 6.4.2) in the automotive domain, many typical characteristics of current in-vehicle communication systems enable unauthorized access and manipulations relatively easy. All information exchange between controllers is done normally completely unencrypted in plain text. Possible messages, their respective structures and communication procedures are specified in freely available documents for most vehicular communication systems. Furthermore, controllers are normally not able to verify, if an incoming message comes from an authorized sender at all. Nevertheless, the major hazard originates from the interconnection of the different vehicular communication systems with each other. The net-spanning data exchange via gateway devices, potentially allows access to any vehicular sub-network from every other existing sub-network. In principle, each vehicular controller is able to send messages to any other existing vehicular controller. Hence, without particular preventive measures, a single comprised vehicular sub-network could endanger the whole in-vehicle communication. In combination with the increasing integration of miscellaneous wireless interfaces and multimedia networks, software programs such as viruses or worms, received over inserted CD/DVDs, email messages, or possibly attached computers, could be able to penetrate highly safety-critical vehicular systems. Even though today gateways already include simple firewall mechanisms, most of them offer unprotected powerful diagnostic functions and interfaces that allow complete access to the vehicular network without any further restrictions.

In the following, some short examples of feasible (integrity) attacks on the protocol layer for each of the representative vehicular bus communication systems introduced afore. Therefor, most in-vehicle communication attacks assume physical or logical access to the corresponding vehicle network via direct contacts to the respective bus wires or by exploiting another (existing or additionally deployed) controller. However, as already pointed out in Section 5.1, most attacks in the automotive domain are internal attacks, which are executed by an authorized or legitimate user of the respective vehicular IT system (*insider*) that hence has physical or logical access to the corresponding vehicle networks.

LIN: Utilizing the dependency of the LIN slaves on their corresponding LIN master, attacking this single point of failure, will be a most promising approach. Introducing well-directed malicious sleep frames deactivates completely the corresponding subnet until a wake-up frame posted by the higher-level CAN bus restores the correct state again. The LIN synchronization mechanism can be another point of attack. Sending frames with bogus synchronization bytes within the SYNCH field makes the local LIN network inoperative or causes at least serious malfunctions. LIN is unprotected against forged messages.

CAN: The priority driven CSMA/CA access control method of CAN network enables attacks that jam the communication channel. Constantly introduced topmost priority nonsense messages will be forwarded always first (even though they will immediately discarded by the receiving controllers) and prevent permanently the transmission of all other CAN messages. Moreover, utilizing the CAN mechanisms for automatic fault localization, malicious CAN frames allow the disconnection of every single controller by posting several well-directed error flags. Furthermore, CAN is vulnerable to forged messages.

FlexRay: Similar to the CAN automatic fault localization, FlexRay's so-called bus guardian can be utilized for the well-directed deactivation of any controllers by appropriate faked error messages. Attacks on the common time base, which would make the FlexRay network completely inoperative, are also feasible, if within one static communication cycle more than f malicious SYNC messages[12] are posted into a FlexRay bus. Moreover, introducing well-directed bogus sleep frames deactivates corresponding power-saving capable FlexRay controllers. FlexRay is also vulnerable to forged messages.

MOST: Since in a MOST network one MOST device handles the role of the timing master, which continuously sends timing frames that allow the timing slaves to synchronize, malicious timing frames are suitable for disturbing or interrupting the MOST synchronization mechanism. Moreover, continuous bogus channel requests, which reduce the remaining bandwidth to a minimum, are a feasible jamming attack on MOST buses. Manipulated false bandwidth statements for the synchronous and asynchronous area within the boundary descriptor of a MOST frame can also make the network completely inoperative. Due to the utilized CSMA/CD access control method used within the asynchronous and the control channel, both are vulnerable to jamming attacks similar to CAN. MOST is also vulnerable to forged messages.

---

[12] $f \geq n/3$, where n is the number of existing FlexRay nodes.

In-Vehicle Communication Security Mechanisms

As already identified at the beginning of this section, in-vehicle communication has to fulfill several security objectives such as message confidentiality or message integrity. In the following, detailed descriptions of three feasible in-vehicle communication security mechanisms are given that together can realize some of the afore identified security objectives, namely message confidentiality, message integrity, and message authenticity. Hence, only validly authenticated and accordingly authorized vehicular controllers are able to read (and send) protected in-vehicle bus messages, while unauthorized modifications can be detected and the origin of a message can be verified. Nevertheless, possible additional objectives on privacy and availability can be realized by appropriate measures at the application layer above (i.e., by the application design) and at the transmission layer below (e.g., by redundancy) respectively. If further freshness of messages is required, this be can be realized, for instance, by securely integrating a real-time clock signal or using cryptographic nonces [13].

Controller Authentication. Authentication of all senders is needed to ensure that only valid controllers are able to communicate within the vehicular bus systems. All unauthorized messages may then be processed separately or are just immediately discarded. Therefore, every controller needs a certificate to authenticate itself against the gateway as a valid sender. A controller certificate $Cert_{ECU}$ consists of the controller identifier ID_{ECU}, the public key PK_{ECU} and the authorizations $Auth_{ECU}$ of the respective controller. The gateway in turn securely holds a list of public keys PK_{OEM} of all accredited OEMs of the respective vehicle. Each controller certificate is digitally signed by the OEM with the corresponding secret key SK_{OEM}. As shown in Table 8.5, the gateway again uses the respective public key of the OEM to verify the validity of the controller certificate. If the authentication process succeeds, the respective controller is added to the gateway's list of validly authenticated controllers, and securely receives its symmetric bus encryption key K_i (cf. subsequent Section).

Encrypted Communication. An essential step to improve the security of vehicular bus communication is the encryption of all in-vehicle data transmissions. Due to the particular constraints of vehicular bus communication systems (computing power, capacity, timing, cf. Section 6.4.1), for instance, a combination of symmetric and asymmetric encryption (*hybrid encryption*) can meet the requirements on

---

[13] A cryptographic nonce is a (random) unpredictable number or bit string used only once by the sender to enable the receiver to verify the freshness of the corresponding message.

Controller Authentication	
1. Verify $(Cert_{ECU}, PK_{OEM})$	Verify ECU's certificate with the corresponding public key of the OEM.
2. Save $(Cert_{ECU})$	Save controller certificate, if the verification succeeds.
3. Encrypt (K_i, PK_{ECU})	Send corresponding symmetric bus encryption key K_i encrypted to the ECU using its public key PK_{ECU}.

Table 8.5: In-vehicle controller authentication protocol based on controller certificates digitally signed by the respective OEM.

sufficient security and high performance. While fast and efficient symmetric cryptography secures the bus-internal broadcast communication, asymmetric cryptography is used to securely handle the distribution of the necessary secret encryption keys. Thus, all controllers of a certain bus system share the same, periodically updated, symmetric encryption key to protect their bus-internal communication. Asymmetric encryption, on the other hand, provides the acquisition of the symmetric key for newly added controllers and carries out the periodic encryption key updates.

In an exemplary realization, as shown in Figure 8.10, a centralized super gateway processor connects all existing vehicular bus systems with each other. Therefore, all across-the-bus communication is done exclusively over this gateway processor. Moreover, the gateway has a protected memory area (cf. Section 8.5) to securely store a set of all symmetric bus encryption keys K_i and the certificates $Cert_j$ of all validly authenticated controllers ECU_j together with their respective authorizations $Auth_j$. In the example, every successfully authenticated vehicular controller ECU_j has received its respective symmetric bus encryption key K_i (cf. Table 8.5) and holds its own public and secret key pair PK_j, SK_j. As all internal bus data is encrypted with K_i, only controllers that hold a valid K_i are able to decrypt and read all locally broadcasted bus messages. Since the centralized gateway further holds all symmetric encryption keys K_i of every bus system connected, secure across-the-bus communication between controller nodes according to their respective authorizations is provided.

As shown in Table 8.6, every controller ECU_j may optionally also receive an individual symmetric authentication key K_j from the central super gateway, to ensure message integrity and to enable sender authentication. If so, each controller could append a message authentication code (MAC) to every message, that means the respective hash value encrypted with its individual authentication key K_j. Even though an asymmetric digital signature scheme could accomplish this task as well

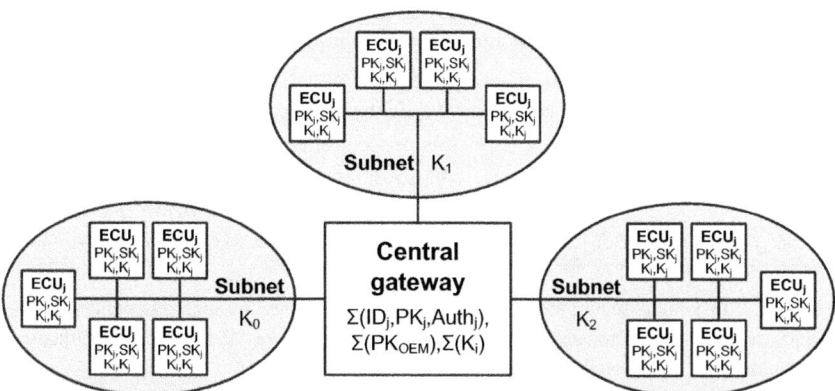

Figure 8.10: Exemplary realization of an encrypted in-vehicle communication using a separate symmetric encryption key for each existing bus system, which has been distributed afore using asymmetric cryptographic schemes.

without additional authentication keys, it would most likely exceed the timing requirements and the computing power of most vehicular controllers.

Controller Message Transmission with Sender Authentication	
1. $C_M = \mathsf{Encrypt}(M, K_i)$	Encrypt message M with bus encryption key K_i.
2. $MAC_M = \mathsf{Encrypt}(\mathsf{Hash}(M), K_j)$	Encrypt hash of M with authentication key K_j.
3. $C = ID_{j'} \|ID_j\|C_M\|MAC_M$	Send C composed of target and sender ECU identity $ID_{j'}$ and ID_j, encrypted message C_M and its message authentication code MAC_M.

Table 8.6: In-vehicle secure message transmission protocol using the shared bus encryption key K_i for message encryption and the individual authentication key K_j to create an optional MAC authentication value.

Table 8.7 shows the receipt of an encrypted message C by a controller or the gateway processor. Whereas network internal controllers decrypt only the symmetric part C_M of C, gateways can verify also the (optionally) enclosed message authentication code MAC_M. Only if the sender authentication succeeds and the sending controller has appropriate authorizations, the gateway re-encrypts and forwards the message across-the-bus into the targeted subnet.

Controller Message Reception and Across-the-bus Forwarding	
1. $M = \text{Decrypt}(C_M, K_i)$	Decrypt $C_M \to M$ with bus encryption key K_i.
2. $\text{Hash}(M) \overset{?}{=} \text{Decrypt}(MAC_M, K_j)$	Verify the integrity of M and authenticate its sender controller using MAC_M (*gateway only*).
3. $\text{Check}(subnet\ i \in Auth_j)$	Check if controller ID_j is allowed to access subnet i based on $Auth_j$ (*gateway only*).
3. $C'_M = \text{Encrypt}(M, K_i)$	Re-encrypt and forward C'_M into target subnet i

Table 8.7: In-vehicle secure message reception protocol using the shared bus encryption key K_i for message decryption and the individual authentication key K_j to verify the authenticity of the message based on the append MAC value.

To enhance the communication security additionally, the gateway may initiate periodic bus encryption key updates. This prevents installing unauthorized controllers by using an afore wiretapped K_i. To inform all controllers of a bus system, for each controller in its current list of validly authenticated controllers, the gateway broadcasts a message encrypted with the respective public key PK_j. When every controller has decrypted its key update message with its secret private key SK_j, a final broadcast of the gateway may activate the new symmetric bus encryption keys.

Gateway Firewalls. For completing internal communication security, vehicular gateways should implement appropriate firewall mechanisms. If the vehicular controllers are capable to implement MACs or digital signatures, the rules of the firewall can be derived from the authorizations given in the certificates of every controller. Therefore, only afore validly authenticated and authorized controllers are able to send messages into (highly safety-relevant) car bus systems. If the vehicular controllers do not have the capabilities to implement MACs or digital signatures, the rules of the firewall can be derived from the authorizations of each vehicular subnet only. However, controllers of lower restricted networks such as LIN or MOST still can be prevented from sending messages into higher safety-relevant bus systems such as CAN or FlexRay. As supplemental security measure, the central gateway should further implement an intrusion detection mechanism. A intrusion detection monitors and analyzes all security-critical activities to detect potential security threats based on anomalies in the vehicular communication behavior. Having detected such an anomaly, which means communication characteristics that differ from known patterns of behavior, the intrusion detection system can raise a warning message (*intrusion alert*) and/or take reasonable general precautions (e.g., disabling a particular controller or service). In case the detected anomaly corresponds also to a known attack pattern (*intrusion signature*), the intrusion detection mechanism can even take active countermeasures (*intrusion response*) that properly thwart the respective attack.

8.6.2 Vehicle-to-Device Communication Security

To notedly enhance vehicular security in general, all garage diagnose functionality as well as all diagnose, test, and verification interfaces, normally used only for inspections in garages or during manufacturing, should be securely disabled by authorized garage personnel to become fully inaccessible during normal operation (cf. Section 5.2.2).

Then, to enable in-vehicle network access for vehicle fault diagnosis or to contact a particular controller (e.g., for software updates or feature activation), the corresponding (diagnosis) device first has to be authenticated successfully at the central gateway controller and/or at the individual ECU to obtain any further access permissions. Again, the application of digital signatures (cf. Section 3.9.1) can help to enable reliable device authentication. Therefore, the vehicle securely [14] holds a set of public keys of trusted external devices (or trusted device manufacturers or trusted OEMs) together with the corresponding access authorizations. Now, once an external device tries to access a controller of the in-vehicle communication network, it first has to present its digital certificate to become verified if the corresponding public key is available. On successful verification, the respective device obtains internal access according to its corresponding authorizations and the in-vehicle firewall rules. Otherwise, the central gateway and/or the respective controller denies any further access.

Mobile passenger devices such as cellular phones, handheld computers, or mobile audio players, which can be interconnected with the vehicle using built-in cable plugs or wireless interfaces (e.g., Bluetooth [Did03] or W-LAN [IEE07]) that cannot or need not to be authenticated, should be able to access only a strictly isolated "multimedia" subnet. The strict isolation from all other in-vehicle communication, for instance by corresponding gateway firewall mechanisms, prevents that unprotected and untrusted external (multimedia) devices can affect critical vehicular communication. Thus, malicious devices connected to a vehicle should at

---

[14] *Securely* in this context means the public keys are at least protected against unauthorized manipulations, substitutions, and fabrications.

most be able to attack the output of existing multimedia displays or stereo speakers[15], but cannot interfere with any critical vehicular application.

8.6.3 Vehicle-to-Infrastructure and Vehicle-to-Vehicle Communication Security

Even though wireless vehicular communications have already been studied in several research projects [Car05, CVI04, Net04, SAF06, SeV06] for a long time, the security issues associated with them have been only hardly tackled. Nonetheless, security of wireless vehicular communication (i.e., V2I, V2V) is now a prominent research area and is largely related to the area of general mobile/ad-hoc communication and network security. Therefore, a vehicle can be simply seen either as a mobile node within a static network similar to a cellular phone, or it can be seen as a short-time wireless node within a mobile, self-organizing ad-hoc network. Hence, many known mobile/ad-hoc communication security mechanisms [EGB02, HJP03, ZH99] can and will be adapted also to the automotive domain to secure vehicular ad-hoc networks (VANET) accordingly. Nevertheless, amongst others according to [ABD⁺06, PP05, RPH06], wireless vehicular communication applications inherently have some characteristics that ease the realization of security measures and others that complicate it. Thus, in addition to the general characteristical advantages and constraints for establishing security in the automotive domain (cf. Section 6.3), the following communication-specific vehicular characteristics can ease the implementation of vehicular communication security further.

Capable Computing Platform. Even though vehicular computing platforms have certain constraints on energy, space, computing resources, and communication bandwidth, in contrast to most other wireless mobile network scenarios (e.g., ad-hoc sensor networks or cellular phones), vehicles usually provide quite adequate capabilities to realize even more sophisticated and somewhat extensive security mechanisms.

Honest majority. As it can be seen for instance with vulnerable pay TV applications, if it continuously requires a considerable effort to circumvent effective security measures, the majority of users will use a reasonably priced application in an honest manner. Moreover, in contrast to pay TV or copy-proof DVDs, most people will shrink away from incalculable vehicular modifications that may affect the safety of the vehicle and hence the safety of the respective occupants. Thus,

---

[15]Nevertheless, the sound volume should still remain only manually adjustable to prevent even malicious "full blast" sounds.

by applying appropriate aggregation schemes, a dishonest minority can usually become outweighed by the honest majority.

Autonomous clock and position data. Since most vehicles will have reliable information regarding their current position and the current time, many security protocols can be enhanced by time and position verifications that complicates several communication attacks (e.g., certain replay and relay attacks).

However, other communication-specific vehicular characteristics can complicate the implementation of vehicular communication security as described in the following.

Highly-dynamic nature. Wireless vehicular communication will be highly dynamic regarding network topologies, effectively involved nodes, and existing connections. Moreover, vehicles usually have only a few seconds to accomplish a particular communication session (i.e, establishment, transmission, and termination). While on the other hand, already delays of seconds could render a message worthless (e.g., in case of an emergency brake warning message).

Large number of different vehicles. Due to the large number of potentially involved vehicles and the various involved manufacturers and divergent effective legislations, security mechanisms for vehicular communication applications have to be highly scalable and particularly compatible to meet the various technical and legal conditions.

Strong authentication versus Privacy. Enabling dependable authentication and enforcing the privacy of the driver at the same time is a known conflict of objectives in many VC applications. For strong dependable authentication, one could employ for instance the string of the vehicle's license plate or even its serial number or the identity of the driver. However, most vehicle users probably will object to VC applications that require them to disclose significantly more private information as usual up to now. Otherwise, current vehicles are neither fully anonymous, but have a unique licence plate attached clearly visible. Thus, successful VC applications must not endanger the privacy of the driver/owner further, but can built up on existing privacy agreements current drivers already accept. This certainly requires a continuous sensitive consideration of any potential practical benefit against possible costs on drivers' privacy.

Open liability issues. Many VC applications suggest or even autonomously conduct crucial driving operations. However, the question who is finally liable for possible consequences in case of a false warning is not yet resolved definitively, but has to be clarified before any VC becomes launched. However, since a small (malicious) error probability likely remains inevitable, VC applications have to implement as many further verification mechanisms and final fall-back solutions as practically feasible.

Conflicting parties. Most VC applications involve several different parties (e.g., vehicle manufacturers, component suppliers, drivers, infrastructure operators, or public authorities) with very different and even conflicting incentives, benefits, and with different portions of the overall costs. Hence, ideally an supranational authority (e.g., an authority of the European Union) has to establish the necessary infrastructure that can be applied independently with prorated costs and even with conflicting interests.

Critical mass. Initially, only a small subset of vehicles will be capable for V2I or V2V communication. High enough penetration rates for practicable V2V applications are for instance expected at the earliest in 2016 [ESE06]. Thus, the chance to meet other VC-enabled vehicles and hence to benefit from the own VC equipment can be rather small. Thus, first VC applications have to provide clear benefits for the respective drivers even if only a minority or specific vehicle groups (e.g., professional and official vehicles such as buses, trucks, taxis, rescue or police vehicles) are equipped with VC mechanisms.

Open Standardization. Even though there already exist several different potential VC application scenarios, there exist no precise and standardized systems or protocols that would allow for interoperable communications and detailed security engineering.

Nevertheless, securing wireless vehicular communication is already the subject of various scientific papers, dedicated research projects [GST05, Net04, SeV06], and workshop series [Int04, Veh05, Veh04]. But, except for the trial-use state IEEE P1609.2 draft standard [IEE06b], which specifies the baselines for developing vehicular networks such as specifications for VANET security services, security message formats as well as corresponding processing and administrative functions, there currently exists virtually no other official VANET security specification or guideline document. And even the IEEE P1609.2 draft does not cover all relevant VANET (security) issues and hence requires lots of further research. Nonetheless, first overall approaches to built VANET security architectures and to secure vehicular communication generally can be found amongst others in [Ger05, GFL+07, HCL04, PGH06, RPH06, RH07]. In addition, the basic underlying security mechanisms are briefly introduced in the following, which can be used as building blocks to thwart most V2I and/or V2V security issues.

Vehicular Key and Certificate Infrastructures

An essential requirement for vehicular communications are appropriate vehicular (public) key and certificate infrastructures that regard the specific VC characteristics described before. This particularly implies the need for an efficient and privacy-preserving issuance of certificates and cryptographic keys by several independent key distribution and certification authorities (CA) as well as the general feasibility and efficiency of certificate revocations for certificates of identified attackers or malfunctioning devices [RPH06, RH07]. In [PBH⁺07] for instance, the authors propose an exemplary vehicular public key infrastructure (VPKI) using a hierarchical structure with cross-certification among individual CAs, whereas [AFWZ07] provides a new VPKI concept, which also provides non-repudiation and enhances privacy. Nevertheless, anonymity should be conditional such that, for instance, in case of law enforcement and liability issues, legitimated authorities can revoke anonymizations [JBP⁺97].

Privacy-Preserving Message Origin Authentication

To determine the relevance and trustworthiness of an incoming VC message, a dependable *and* privacy-preserving message origin authentication is essential (cf. previous paragraph). In general, this means rather the trustworthy verification of the sender's authorization (i.e., the sender is properly legitimated and not an arbitrary radio station) and the corresponding classification (e.g., as an official or regular vehicle, as a road sign or other authorized road side units) rather than the authentication of a globally unique and permanent identifier such as the vehicle's serial number. In [CPHL07, RH07], several efficient and privacy-preserving vehicular authentication schemes are proposed, which provide non-repudiation and prevent impersonation based on robust pseudonyms. Similar approaches deploy dynamic short-time pseudonyms [GG07, PBH⁺07], anonymization services [PP05], or anonymous broadcasting [LB07].

Secure Positioning

Secure positioning regards the individual secure localization on the one hand and on the other hand a trustworthy and privacy-preserving provision of one's own position information to others to prevent for instance relay attacks. Usual GPS signals for example are relatively imprecise and can be forged and jammed easily [WJ03]. Adding a fully tamper-protected GPS receiver could be a possible solution. However, adding tamper-protection is costly and has to take sophisticated hardware attacks into account (cf. Section 7.1). Adding asymmetric security mechanisms could be another solution for position security [Kuh04], but is hardly practicable with today's common satellite-based navigation systems. Without tamper-protected positioning, localization schemes based on distance-bounding and verifiable multilateration [HCL04] can assure that a vehicle is inside of a well-defined geographical area.

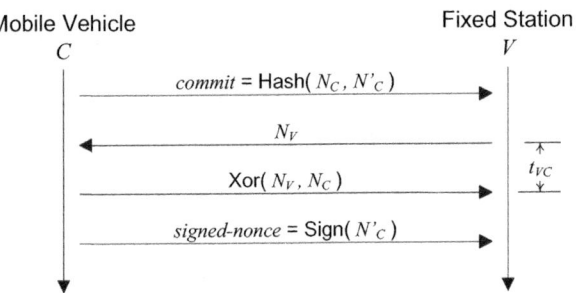

Figure 8.11: Exemplary distance-bounding protocol according to [HCL04].

Figure 8.11 shows a distance-bounding protocol between a mobile vehicle C (*claimer*) and a fixed base station V (*verifier*) using a mutually authenticated communication channel as proposed by [HCL04]. There, the vehicle C firstly commits the hash value of two self-generated random nonces N_C and N'_C. The verifier V then sends a challenge nonce N_V, which C immediately responds with the XOR of $N_V \otimes N_C$. The verifier V now can calculate the *minimal* distance d_{min} between V and C based on the measured challenge-response time t_{VC} and the speed of light constant c such that $d_{min} = c * t_{VC}$. Thus, C cannot send $N_V \otimes N_C$ before it has received N_V, however, C can arbitrarily delay the response, thus pretending to be farther from V as it actually is. Lastly, C signs N'_C from its initial commitment using its public key material and sends *sig-nonce* to V to prove that *commit* belongs to C's response. By deploying multiple variedly located verifiers and methods of triangulation, the actual location of V can be sufficiently verified, because if C would maliciously increase the measured distance to one verifier V_1, C would have to decrease the distance to another verifier V_2 accordingly to maintain plausibility. However, pretending to be more closer to V_2 is actually impossible, since it would require C to response to V_2 faster than the speed of light.

In [LSK06] the authors propose detection mechanisms that are capable of recognizing nodes cheating about their location in position beacons that do not rely on special hardware or dedicated infrastructure. Other related approaches provide a secure localization of a vehicle's relative position based on the broadcasts of surrounding vehicles [PP05] or focus on location privacy [SHL$^+$05].

Communication Integrity

For vehicular communication integrity, existing network security schemes can quite readily be translated to meet the typical vehicular constraints particularly regarding efficiency and complexity as well. Appropriate approaches to detect and correct malicious data in VANETs are described in [GGS04]. Communication integrity in a broader sense comprises also the detection of logically false information for instance by applying plausibility checks. Appropriate schemes based on message aggregation and group communication [RAH06], which in turn rely on an honest majority, thus can help to increase the dependability of VC information further.

9 Organizational Security

This chapter provides an introduction into organizational security aspects in the automotive domain. It covers general challenges for organizational security in vehicular engineering and some specific organizational security aspects during the typical lifecycle of a vehicle or a vehicular component.

9.1 The Safety of Secrets

Vehicular IT security always comprises both technical *and* organizational measures. The lifecycle of a usual vehicle or vehicular component always includes various security-critical organizational structures, processes, and policies as well. In some cases, a (vehicular) IT system can be compromised only due to some organizational weaknesses (e.g., by social engineering or just by carelessness). Thus, often already some small organizational failures can lead to devastating consequences [Pri08]. Besides the considerable financial losses and the potential losses of expertise [CMR04], information security breaches that become publicly known can furthermore considerably damage the corporate's reputation as well as the customers' trust [Cam03]. Thus, to maintain the protection of vehicular IT systems and to protect the critical assets of a (vehicular) company, dependable organizational structures, secure organizational processes, and appropriate security policies are mandatory. Organizational security aspects [1] cover the complete lifecycle of a vehicle or a vehicular component, which includes vehicular research and development, vehicular manufacturing, as well as corresponding vehicle service and maintenance. Because of their particular importance for vehicular IT security, respective organizational security measures have to be considered individually and additionally to all the technical measures treated in this work. Organizational security, furthermore, normally is an inherent part of nearly all IT security evaluations [ISO05a]. Anyhow, further readings on this subject are helpful and can be found amongst others in [Bun04, Bun06, ISO04a, ISO05c, ISO07b, JG07].

---

[1] Note, this interpretation of organizational security refers to the protection against different information security breaches and does not treat organizational security measures to protect a company's critical business functions in case of a "conventional disaster" such as a sudden loss of resources or personnel on a large scale.

Despite its pivotal importance for most manufacturers and suppliers in the automotive domain, organizational security is often still neglected. Hence, the major challenges in "embedding information security into [vehicular] organizations" [JG07] and to ensure "the safety of [vehicular] secrets" [Joh04] amongst others are:

Understanding reasons and proper application. Establishing (or improving) organizational security in general first of all means establishing a sound understanding for the need of organizational security measures. This includes raising an awareness for possible vulnerabilities and risks as well as providing education for proper application *and* understanding of available security services for all employees throughout the hierarchies. Only if employees understand the underlying reasons and the correct application of all organizational security measures, they will effectively change their behavior and integrate and sustain security into the organizational culture of the company.

Realistic and enforceable policies. Ensuring proper information security while keeping the actual operating procedures as simple as possible can quickly become a conflicting target. Hence, to maintain security while not too much interfering with the actual operating procedures, all security policies have to be balanced carefully. Thus, organizational security policies have to be realistic and enforceable, even though security professionals often have only fewer understanding of the actual business and vice versa.

Multitude of involved parties. As already mentioned in Section 6.4.2, the current multi-tier vehicular manufacturing process (OEM and possibly several layers of suppliers) considerably increases the complexity and the costs of organizational security measures. The heavy use of outsourcing, subcontracting, and various partnerships, as fairly typical in the automotive domain, involves many external parties with very different security understandings, security policies, and hence very different levels of (mutual) trustworthiness. Thus, it can be very difficult to establish a minimal holistic security policy to prevent that the collective security level is limited to the security level of the weakest party that is involved.

Multitude of involved regulations and cultures. Up to the Internet age, most vehicular companies had employees of similar culture working in a few manageable locations. Today the workplaces of a single company already are spread all other world, hence involving many different laws, regulations, standards, and cross-cultural differences regarding the level of security and privacy for and to employees, partners, and customers. Establishing holistic security policies that comply with all effective laws, regulations, and standards while adequately regarding all cultural differences as well will be very difficult, if not impossible at all.

Costs with hardly quantifiable benefits. As with most security measures (cf. Section 6.4.2), most organizational security measures also cause costs, consume time, and require personnel without providing immediately apparent benefits. Even though there emerge first models to measure and quantify the returns on security investments [Son06], most organizational security measures lack of appropriate metrics to measure their power for instance in terms of prevented losses or enhanced trustworthiness. Thus, without any clearly quantifiable benefits, most organizational security measures usually have to get by with only limited personnel and only limited financial resources.

9.2 Achieving Organizational Security in the Automotive Domain

Nevertheless, most vehicular companies are at least aware that they require a standard security infrastructure for everyday processes such as secure communication (e.g. by e-mail encryption and e-mail signatures), an appropriate access control system to all internal information and resources, a secure network including firewalls, anti-virus solutions, intrusion detection systems, and so on. Furthermore, physical access to sensitive information and resources should be restricted to the necessary minimum of employees (e.g., physical access to the respective facilities of a company should be restricted to the employees that actually are working there). Unfortunately, there are virtually no general-purpose advises to ensure holistic organizational security. Thus, organizational security measures need to be developed and evaluated individually in order to properly regard the individual environment, existing processes and individual (security) objectives. However, in the following at least some general guidelines for managing organizational security from the vehicular manufacturer's perspective are given.

(1) All critical assets of a vehicular company (e.g., cryptographic secrets, product specifications, intellectual property, business information, or trade secrets) should be identified and explicitly described together with all corresponding organizational security measures currently effective in order to protect them. This includes all effective technical measures, corresponding organizational processes and procedures, and all effective security policies.

(2) All organizational security measures should be evaluated based on potential attacks and possible vulnerabilities. This should be done only by skilled security experts, which ideally also have an automotive background.

(3) All security weaknesses that were identified as result of the evaluation should be categorized according to their individual attack potential and their potential damage in order to compile a meaningful risk evaluation.

(4) All critical security weaknesses (according to the previous risk evaluation) should be eliminated. Whereas, all correspondingly newly developed organizational security measures should be evaluated again to verify their actual effectiveness.

(5) All effective organizational security measures should be revised routinely (e.g., annually) to become continuously adopted to potentially changed conditions, particularly, if new processes or products are introduced.

Lastly, an ISO 27001 [ISO07b] certification, for instance, can provide at least some basic level of assurance for having a proper information security management implemented. And even though it cannot guarantee completeness, it at least recommends some helpful practices for proper information security management in organizations that comprise amongst others:

- Risk assessment and treatment,

- Security policy definition,

- Organizational security structures, reporting, and liaison,

- Asset management,

- Human resources security,

- Physical and environmental security,

- Communications and operations management,

- Access control,

- Information systems acquisition, development and maintenance,

- Information security incident management,

- Business continuity management, and

- Compliance with legal requirements, security policies and standards, technical standards, and security audits.

9.3 Organizational Security Measures in a Vehicular Lifecycle

Organizational security measures should be integrated from the outset of every vehicular development, since this approach turned out to be considerably less costly and less time-consuming in comparison to the still widespread approach of trying to add organizational security measures first when the basic necessity arises (which nevertheless could be too late at all). Within the automotive domain several different phases have to be considered in order to ensure organizational security for the complete lifecycle of a vehicle or a vehicular component. Hence, the main phases of a vehicular lifecycle that have to be covered at least are (i) vehicular research and development, (ii) vehicular manufacturing, and (iii) vehicle service and maintenance. In the following, several important organizational security aspects for each phase of such a typical vehicular lifecycle are described.

9.3.1 Research and Development

During vehicular research and development (R&D) the most important security-critical organizational processes are as follows:

System design. The design of a new vehicular IT system involves lots of expertise, working hours, financial capital, or even trade secrets (cf. Section 4.3). Thus, a (new) design should be regarded as strictly confidential and should not be published—even within the organization—to avoid, for instance, typical insider attacks. For all information security management aspects, corresponding security policies have to developed and verified accordingly (cf. Section 9.2). This regards particularly the confidential collaboration with external entities that are potentially involved.

Implementation. The implementation often summarizes the respective expertise, and again, could include important trade secrets and valuable intellectual property. Hence, the implementation should be split up into several individual subtasks in such a way that the respective developers have only access to a small subset of the actual information, which they implicitly require for their respective task. Furthermore, any implementation results (e.g., program code, documents, or hardware layouts) always have to be handled confidentially and verified for integrity. In particular, it needs to be ensured that a developer cannot insert any vulnerabilities or hidden backdoors (e.g., for undocumented testing purposes). Thus, independent peer reviews of all implementations should be mandatory for ensuring safety and security. Lastly, a secure version control system should be applied that keeps track of all work and all changes to the corresponding implementation(s).

Testing. Often testing or pre-release versions of hardware and in particular software components leak unauthorizedly into public, which for instance almost completely reveal underlying designs to competitors or counterfeiters. Thus, strict security policies and access control mechanisms should be mandatory *particularly* for testing purposes.

9.3.2 Manufacturing

In general, the manufacturing environment in the automotive domain should not be considered as a secure or trustworthy environment. A manufacturing environment usually involves a multitude of different people, whose trustworthiness can be neither verified nor supervised completely. Moreover, it is seldom possible to strictly enforce the respective (physical) access restrictions for all production facilities that are involved. Lastly, securing a manufacturing environment is costly, laborious, and error-prone. Thus, it is highly recommended to design vehicular components that do not rely on a secure production environment or whose security-critical production step (e.g., the injection of a secret cryptographic key) at least can completely be done within a small enclosed area ideally at the end of the manufacturing process.

9.3.3 Service and Maintenance

Insider attacks are an important organizational security issue, not only in the automotive domain. The service and maintenance personnel, for example, usually has a deep knowledge of most employed vehicular IT systems and access to necessary equipment. It is also known that several car dealers and mechanics sell up their knowledge and corresponding skills to manipulate the vehicular control units of their customers (cf. Chapter 5). Hence, during service and maintenance it is virtually impossible to provide a secure environment where the involved personnel can be assumed to be trustworthy. Quite the contrary, motorcar mechanics even have a considerably above-average portion of people among with a criminal background [And03]. Hence, from the very beginning it should be assumed that processes in the service area are performed in an insecure environment by people that are not trustworthy. Hence, the rights of service personnel should be heavily restricted. Any access to security relevant methods should be securely logged (cf. Section 4.6.5) and checked for plausibility in order to be able to reconstruct any critical procedures.

10 Conclusions

By now, security engineering is an accepted challenge in the development of most vehicular electronics. Nevertheless, vehicular security engineering is still a recent field of research and an emerging area of application. However, the ongoing demand for more advanced and comprehensive vehicular applications that more and more have direct impacts on crucial driving operations impose strong obligations on trustworthiness, reliability, dependability, and hence also on security. Unfortunately, there exist no ready-made standard solutions yet that can be simply added to vehicular applications to ensure necessary IT security. In fact, securing vehicular electronics usually requires very individual and well-adapted approaches that seldom can simply be derived from existing conventional IT protection measures. Vehicular security measures have to be considered already at the beginning of every development process and require both technical and organizational measures. However, in doing so, embedding IT security into vehicles then:

(1) protects against unauthorized manipulations by external *and* internal attackers,

(2) increases the safety, reliability, and dependability of a vehicular system,

(3) ensures the requirements on trustworthiness and privacy of drivers and passengers,

(4) protects liability, revenues, and expertise of vehicle manufacturers and suppliers,

(5) will be a necessary requirement for many vehicular IT applications, but also

(6) enables a multitude of new vehicular legacy and business applications.

However, as already described in Section 6.4, there are several different difficulties to overcome in order to develop strong vehicular security solutions. Vehicular IT security measures have to deal with very specific boundary conditions regarding available computing resources, the physical environment, cost requirements, and maintainability. They further have to meet several non-technical and organizational challenges. IT security solutions have not least to be designed extremely carefully. A single "minor" flaw in the system design can render the entire solution insecure. This is quite different from engineering most other technical systems, where a single non-optimum component usually does not invalidate the entire system. Lastly, security and cryptography has historically been a field dominated by theoreticians, whereas vehicular IT is usually done by engineers. The culture in those two communities is quite different at times, and both sides have to put effort into developing mutual expertises and understanding each other's way of thinking and communicating.

Bibliography

[ABD⁺06]   Amer Aijaz, Bernd Bochow, Florian Dötzer, Andreas Festag, Matthias Gerlach, Rainer Kroh, and Tim Leinmüller. Attacks on Inter-Vehicle Communication Systems – An Analysis. In *WIT 2006: Workshop on Intelligent Transportation*, 2006.

[AES⁺07]   N. Asokan, Jan-Erik Ekberg, Ahmad-Reza Sadeghi, Christian Stüble, and Marko Wolf. Enabling Fairer Digital Rights Management with Trusted Computing. In *Proceedings of the 10th International Conference on Information Security, ISC 2007, Valparaiso, Chile, October 9 – 12, 2007*. Springer-Verlag, 2007.

[AF04]   Tiago Alves and Don Felton. TrustZone: Integrated Hardware and Software Security. White paper, Advanced RISC Machines Limited (ARM), July 2004.

[AFWZ07]   Frederik Armknecht, Andreas Festag, Dirk Westhoff, and Ke Zeng. Cross-layer Privacy Enhancement and Non-repudiation in Vehicular Communication. In *4th Workshop on Mobile Ad-Hoc Networks (WMAN)*, 2007.

[AHS05]   André Adelsbach, Ulrich Huber, and Ahmad-Reza Sadeghi. Secure Software Delivery and Installation in Embedded Systems. In *Information Security Practice and Experience Conference, ISPEC 2005, Singapore, April 11-14, 2005*, volume 3439 of *LNCS*, pages 255–267. Springer-Verlag, 2005.

[AHSS05]   André Adelsbach, Ulrich Huber, Ahmad-Reza Sadeghi, and Christian Stüble. Embedding Trust into Cars - Secure Software Delivery and Installation. In *3rd Workshop on Embedded Security in Cars, escar 2005, Bochum, Germany, November 29 – 30*, 2005.

[Ame04]   Sandro Amendola. Improving Automotive Security by Evaluation – From Security Health Check to Common Criteria. White paper, Security Research & Consulting GmbH, 2004.

[And98]   Ross Anderson.   On the Security of Digital Tachographs.   In
             *ESORICS '98: Proceedings of the 5th European Symposium on
             Research in Computer Security*, pages 111–125. Springer-Verlag,
             1998.

[And01]   Ross Anderson. *Security Engineering: A Guide to Building Depend-
             able Distributed Systems*. John Wiley & Sons, Inc., New York, NY,
             USA, 2001.

[And03]   Ross Anderson. Electronic Safety and Security – New Challenges
             for the Car Industry. In *1st Workshop on Embedded Security in Cars,
             escar 2003, Bochum, Germany, November 18 – 19*, 2003.

[ANS95a]   ANSI X9.17:1995. *Financial Institution Key Management (Whole-
             sale)*. American National Standard Institute, 1995.

[ANS95b]   ANSI X9.62:2005.   *Public Key Cryptography for the Financial
             Services Industry, The Elliptic Curve Digital Signature Algorithm
             (ECDSA)*. American National Standard Institute, 1995.

[AOS+08]   N. Asokan, André Osterhues, Ahmad-Reza Sadeghi, Christian
             Stüble, and Marko Wolf. Securing Peer-to-peer Distributions for
             Mobile Devices. In *Proceedings of the 4th Information Security
             Practice and Experience Conference, ISPEC 2008, Sydney, Aus-
             tralia, April 21 – 23, 2008*. Springer-Verlag, 2008.

[Atm08]   Atmel Corporation. Secure Microcontrollers, Smartcard and Secu-
             rity ICs. *www.atmel.com/products/*, 2008.

[Aud08]   Audi AG. *www.audi.com*, 2008.

[Aut03]   The Automotive Open System Architecture (AUTOSAR) Develop-
             ment Partnership. *www.autosar.org*, 2003.

[Aut04]   The Automotive Safety & Security Workshop Series.   *www.
             automotive2004.de*, 2004.

[Ava08]   Avaya Inc. Research Labs. LibSafe – Protecting Critical Elements
             of Stacks. *www.research.avayalabs.com*, 2008.

[Bac97]   Anton Bachhuber. Immobilizer for Preventing Unauthorized Start-
             ing of a Motor Vehicle and Method for Operating the Same. US
             Patent 5,675,490, October 1997.

[Ban03]   The Bangkok Post. Computer Traps Thailand's Finance Minister
             Suchart. The Bangkok Post, May 13, 2003.

[BCC04]   Ernie Brickell, Jan Camenisch, and Liqun Chen. Direct Anonymous
             Attestation. In *Proceedings of the 11th ACM Conference on Com-
             puter and Communications Security, CCS 2004, Washington DC,
             USA, October 25 – 29, 2004*. ACM Press, 2004.

[BCG⁺07]   Neil Bird, Claudine Conrado, Jorge Guajardo, Stefan Maubach,
             Geert Jan Schrijen, Boris Skoric, Anton M. H. Tombeur, Peter
             Thueringer, and Pim Tuyls. ALGSICS – Combining Physics and
             Cryptography to Enhance Security and Privacy in RFID Systems.
             In *4th European Workshop on Security and Privacy in Ad-Hoc and
             Sensor Networks (ESAS 2007)*, volume 4572 of *LNCS*, pages 187–
             202. Springer-Verlag, 2007.

[BDF⁺03]   Paul Barham, Boris Dragovic, Keir Fraser, Steven Hand, Tim Har-
             ris, Alex Ho, Rolf Neugebauer, Ian Pratt, and Andrew Warfield. Xen
             and the Art of Virtualization. In *SOSP '03: Proceedings of the
             nineteenth ACM symposium on Operating systems principles*, pages
             164–177. ACM Press, 2003.

[BDI⁺07]   Eli Biham, Orr Dunkelman, Sebastiaan Indesteege, Nathan Keller,
             and Bart Preneel. How to Steal Cars – A Practical Attack on KeeLoq.
             Rump Session at CRYPTO '07, August 2007.

[BE04]   Jeremy Blum and Azim Eskandarian. The Threat of Intelligent Col-
             lisions. *IEEE IT Professional*, 6(1):24–29, 2004.

[BEPW07]   Andrey Bogdanov, Thomas Eisenbarth, Christof Paar, and Marko
             Wolf. *Trusted Computing*, chapter Trusted Computing in Automo-
             tive Systems. Vieweg-Verlag, 2007.

[BEWW07]   Andrey Bogdanov, Thomas Eisenbarth, Marko Wolf, and Thomas
             Wollinger. Trusted Computing for Automotive Systems – New Ap-
             proaches to Enforce Security for Electronic Systems in Vehicles.
             In *23th VDI/VW-Gemeinschaftstagung: Automotive Security, Wolfs-
             burg, Germany, November 28 – 29, 2007*. VDI-Verlag, 2007.

[BFM⁺03]   M. Baleani, A. Ferrari, L. Mangeruca, A. Sangiovanni-Vincentelli,
             Maurizio Peri, and Saverio Pezzini. Fault-Tolerant Platforms for Au-
             tomotive Safety-Critical Applications. In *CASES '03: Proceedings*

of the 2003 international conference on Compilers, architecture and synthesis for embedded systems, pages 170–177. ACM Press, 2003.

[Bri07]   The British Broadcasting Corporation (BBC) News United Kingdom. Drivers Stranded by Car Signals. *http://news.bbc.co.uk/2/hi/uk_news/england/kent/7073935.stm*, November 1, 2007.

[Bro04]   Manfred Broy. Sichere Software im Automobil – Potenziale, Herausforderungen,Trends. In *2nd Workshop on Embedded Security in Cars, escar 2004, Bochum, Germany, November 10 – 11*, 2004.

[Bro06]   Manfred Broy. Challenges in Automotive Software Engineering. In *ICSE '06: Proceedings of the 28th International Conference on Software Engineering*, pages 33–42. ACM Press, 2006.

[Bug08]   The BugTraq Mailing List. *www.securityfocus.com*, 2008.

[Bun04]   Bundesamt für Sicherheit in der Informationstechnik (BSI). IT Baseline Protection Manual. *www.bsi.de/english/gshb/*, 2004.

[Bun06]   Bundesamt für Sicherheit in der Informationstechnik (BSI). IT-Grundschutz Kataloge. *www.bsi.de/gshb/*, 2006.

[Cam03]   K. Campbell. The Economic Cost of Publicly Announced Information Security Breaches: Empirical Evidence from the Stock Market. *Journal of Computer Security*, 11(3):431–448, 2003.

[Car05]   Car-2-Car Communication Consortium. *www.car-2-car.org*, 2005.

[CCA02]   Erik Coelingh, Pascal Chaumette, and Mats Andersson. Open-Interface Definitions for Automotive Systems – Application to a Brake-By-Wire System. Technical Paper 2002-01-0267, SAE International, March 2002.

[CCD07]   CCDB-2007-04-001. *Application of Attack Potential to Smartcards, Common Criteria Supporting Document and Mandatory Technical Document Version 2.3*. Netherlands National Communications Security Agency (NLNCSA), 2007.

[CH06]   Srdjan Capkun and Jean-Pierre Hubaux. Secure Positioning in Wireless Networks. *IEEE Journal on Selected Areas in Communications: Special Issue on Security in Wireless Ad Hoc Networks*, 24(2):221–232, 2006.

[CMR04]   H. Cavusoglu, B. Mishra, and S. Raghunathan. The Effect of In-
           ternet Security Breach Announcements on Market Value: Capital
           Market Reactions for Breached Firms and Internet Security Devel-
           opers. *International Journal of Electronic Commerce*, 9(1):70–104,
           2004.

[CNN07]   CNNMoney.com Business Website. Fake Parts Reportedly Cost
           Ford $1B. *http://money.cnn.com/2007/01/22/news/companies/
           ford_counterfeit_parts/*, January 22, 2007.

[CPHL07]   Giorgio Calandriello, Panos Papadimitratos, Jean-Pierre Hubaux,
           and Antonio Lioy. Efficient and Robust Pseudonymous Authentica-
           tion in VANET. In *VANET '07: Proceedings of the 4th International
           Workshop on Vehicular Ad-hoc Networks*, pages 19–28. ACM Press,
           2007.

[CTG03]   Fulvio Corno, S. Tosato, and P. Gabrielli. System-Level Analysis of
           Fault Effects in an Automotive Environment. In *18th IEEE Inter-
           national Symposium on Defect and Fault Tolerance in VLSI Systems
           (DFT'03)*, pages 529–536. IEEE Press, 2003.

[CVI04]   CVIS: The Cooperative Vehicle-Infrastructure Systems Project.
           www.cvisproject.org, 2004.

[CW07]   Brian Chess and Jacob West. *Secure Programming with Static Anal-
           ysis*. Addison-Wesley, 2007.

[DD06]   Driver and Vehicle Licensing Agency (DVLA). Electronic Num-
           ber Plates (ENP). Feasibility Report Version 3.0, United Kingdom
           Department for Transport, *www.dvla.gov.uk/media/pdf/other/enp_
           report.pdf*, December 2006.

[Des94]   Yvo G. Desmedt. Threshold Cryptography. *European Transactions
           on Telecommunications*, 5(4):449–457, 1994.

[Deu08]   Deutscher Kraftfahrzeug-Überwachungsverein (DEKRA). *www.
           dekra.de*, 2008.

[DGL+02]   S. Duri, M. Gruteser, X. Liu, P. Moskowitz, R. Perez, M. Singh,
           and J.M. Tang. Framework for Security and Privacy in Automo-
           tive Telematics. *Proceedings of the 2nd International Workshop on
           Mobile Commerce*, pages 25–32, 2002.

[DH76]   Whitfield Diffie and Martin E. Hellman. New Directions in Cryp-
            tography. *IEEE Transactions on Information Theory*, IT-22(6):644–
            654, 1976.

[Día05]   Claudia Díaz. Anonymity Metrics Revisited. In *Dagstuhl Seminar
            on Anonymous Communication and its Applications*, 2005.

[Did03]   Myra Dideles. Bluetooth: A Technical Overview. *ACM Crossroads*,
            9(4):11–18, 2003.

[DoD85]   DoD 5200.28-STD. *Trusted Computer System Evaluation Criteria
            (TCSEC)*. United States Department of Defense (DoD), 1985.

[DOT04]   DOT HS 809 859. *Vehicle Safety Communications Project Task 3 Fi-
            nal Report, Identify Intelligent Vehicle Safety Applications Enabled
            by DSRC*. United States Department of Transportation – National
            Highway Traffic Safety Administration, 2004.

[DR98]   Joan Daemen and Vincent Rijmen. The Block Cipher Rijndael. In
            *Smart Card Research and Applications Conference, CARDIS 1998,
            Louvain-la-Neuve, Belgium, September 14 – 16, 1998*, volume 1820
            of *LNCS*, pages 277–284. Springer-Verlag, 1998.

[Dri02]   Kevin R. Driscoll. Safety in Automotive Industry. TTA-Group Fo-
            rum, Munich, Germany, November 15, 2002.

[e-P08]   e-Plate Limited. *www.e-plate.com*, 2008.

[E+08]   David Evans et al. Splint – Annotation-Assisted Lightweight Static
            Code Checking. *www.splint.org*, 2008.

[EAM03]   A. de la Escalera, J. M. Armingol, and M. Mata. Traffic Sign Recog-
            nition and Analysis for Intelligent Vehicles. *Image and Vision Com-
            puting*, 21(3):247–258, March 2003.

[EGB02]   Laurent Eschenauer, Virgil Gligor, and John Baras. On Trust Estab-
            lishment in Mobile Ad-Hoc Networks. In *Security Protocols: 10th
            International Workshop, Cambridge, UK, April 17 – 19*, 2002.

[EGP+07]   Thomas Eisenbarth, Tim Güneysu, Christof Paar, Ahmad-Reza
            Sadeghi, and Marko Wolf. Reconfigurable Trusted Computing in
            Hardware. In *Proceedings of the 2nd Workshop on Scalable Trusted
            Computing, STC 2007, Alexandria, Virginia, USA, November 2,
            2007*. ACM Press, 2007.

[EHH⁺05]   Kevin Elphinstone, Gernot Heiser, Ralf Huuck, Stefan M. Petters, and Sergio Ruocco. L4Cars. In *3rd Workshop on Embedded Security in Cars, escar 2005, Bochum, Germany, November 29 – 30*, 2005.

[Ehl03]   Tiemo Ehlers. Systemintegrität von vernetzter Fahrzeugelektronik. In *1st Workshop on Embedded Security in Cars, escar 2003, Bochum, Germany, November 18 – 19*, 2003.

[EKP⁺07]   Thomas Eisenbarth, Sandeep Kumar, Christof Paar, Axel Poschmann, and Leif Uhsadel. A Survey of Lightweight Cryptography Implementations. *IEEE Design & Test of Computers – Special Issue on Secure ICs for Secure Embedded Computing*, 24(6):522–533, 2007.

[ElG85]   Taher ElGamal. A Public Key Cryptosystem and a Signature Scheme Based on Discrete Logarithms. *IEEE Transactions on Information Theory*, 31(4):469–472, 1985.

[EM00]   Richard Evans and Jonathan D. Moffett. Derivation of Safety Targets for the Random Failure of Programmable Vehicle Based Systems. In *19th International Conference on Computer Safety, Reliability and Security, SAFECOMP 2000, Rotterdam, The Netherlands, October 24 – 27, 2000*, volume 1943 of *LNCS*, pages 240–249. Springer-Verlag, 2000.

[Emb03]   The Embedded Security in Cars (escar) Workshop Series. *www.escarworkshop.org*, 2003.

[EN 04]   EN 12253:2004. *Road Transport and Traffic Telematics. Dedicated Short-range Communication. Physical Layer Using Microwave at 5.8 GHz*. European Committee for Standardization (CEN), July 2004.

[ER03]   Ralf Engers and Bernd Rössler. TPM, spezifiziert durch TCG/TCPA – eine potenzielle Lösung für Sicherheitsfragen in der Automobilindustrie. In *1st Workshop on Embedded Security in Cars, escar 2003, Bochum, Germany, November 18 – 19*, 2003.

[eSa08]   eSafety Security Working Group. *www.esafetysupport.org*, 2008.

[ESE06]   Stephan Eichler, Christoph Schroth, and Jörg Eberspächer. Car-to-Car Communication. In *Proceedings of the VDE Kongress — Innovations for Europe*, 2006.

[Eur02]   European Commission Regulation (EC) No 1360/2002. *Corri-gendum to Commission Regulation Adapting for the Seventh Time to Technical Progress Council Regulation (EEC) No 3821/85 on Recording Equipment in Road Transport.* European Commission, June 2002.

[Eur06]   European Commission Regulation (EC) No 561/2006. *Commission Regulation on the Harmonisation of Certain Social Legislation Relating to Road Transport and Amending Council Regulations (EEC) No 3821/85 and (EC) No 2135/98 and Repealing Council Regulation (EEC) No 3820/85.* European Commission, March 2006.

[Eur08]   European Emergency Call Driving Group (eCall). *http://europa.eu. int/information_society/activities/esafety/forum/ecall/*, 2008.

[EZMTV02]   M. El Zarki, S. Mehrotra, G. Tsudik, and N. Venkatasubramanian. Security Issues in a Future Vehicular Network. *European Wireless*, February 2002.

[Fib04]   Lenka Fibikova. On Building Trusted Services in Automotive Systems. In *2nd Workshop on Embedded Security in Cars, escar 2004, Bochum, Germany, November 10 – 11*, 2004.

[FIP77]   FIPS-46-3. *Data Encryption Standard (DES).* National Institute of Standards and Technology, 1977. Reaffirmed in October 1999.

[FIP94]   FIPS-186-3. *Digital Signature Standard (DSS).* National Institute of Standards and Technology, 1994. Reaffirmed in March 2006.

[FIP01]   FIPS-197. *Advanced Encryption Standard (AES).* National Institute of Standards and Technology, 2001.

[FIP02a]   FIPS-140-2. *Security Requirements for Cryptographic Modules.* National Institute of Standards and Technology, 2002.

[FIP02b]   FIPS-180-2. *Secure Hash Standard.* National Institute of Standards and Technology, 2002.

[FIP02c]   FIPS-198. *The Keyed-Hash Message Authentication Code (HMAC).* National Institute of Standards and Technology, 2002.

[FL06]   Igor Furgel and Kerstin Lemke. A Review of the Digital Tachograph System. In *Embedded Security in Cars: Securing Current and Future Automotive IT Applications*, pages 69–94. Springer-Verlag, 2006.

[Fle00]   The FlexRay Consortium. *www.flexray.com*, 2000.

[For08]   Fortify Inc. RATS – Rough Auditing Tool for Security. *www.fortifysoftware.com/security-resources/rats.jsp*, 2008.

[Fra08]   Fraunhofer Institute for Photonic Microsystems (Fraunhofer IPMS). IPMS_RSA – A Core for RSA Algorithm. *www.ipms.fraunhofer.de/en/products/sensor_cores.shtml*, 2008.

[Fre08]   Free Software Foundation Inc. gprof – The GNU Profiler. *www.gnu.org*, 2008.

[Fri93]   William F. Friedman. *Military Cryptanalysis: Part I–IV*. Aegean Park Press, 1993.

[Fri04]   Hans-Georg Frischkorn. Automotive Software – The Silent Revolution. In *Workshop on Future Generation Software Architectures in the Automotive Domain, San Diego, CA, USA, January 10 – 12*, 2004.

[FW07a]   Martin Feldhofer and Johannes Wolkerstorfer. Strong Crypto for RFID Tags – A Comparison of Low-Power Hardware Implementations. In *IEEE International Symposium on Circuits and Systems (ISCAS)*, pages 1839–1842. IEEE Press, 2007.

[FW07b]   Franz Fürbass and Johannes Wolkerstorfer. ECC Processor with Low Die Size for RFID Applications. In *IEEE International Symposium on Circuits and Systems (ISCAS)*, pages 1835–1838. IEEE Press, 2007.

[Gal98]   Peter Galvin. Designing Secure Software. SunWorld, 1998.

[Gar03]   Ed Garsten. Fake parts hobble car industry. Detroit News, October 2, 2003.

[GCvDD02]   Blaise Gassend, Dwaine Clarke, Marten van Dijk, and Srinivas Devadas. Silicon Physical Random Functions. In *Proceedings of the 9th ACM Conference on Computer and Communications Security*, pages 148–160. ACM Press, 2002.

[Ger05]   Matthias Gerlach. VaneSe – An Approach to VANET Security. In *Workshop on Vehicle to Vehicle Communications (V2VCOM), San Diego, California, USA, July 21*, 2005.

[GFL⁺07]   Matthias Gerlach, Andreas Festag, Tim Leinmüller, Gabriele Goldacker, and Charles Harsch. Security Architecture for Vehicular Communication. In *WIT 2007: Workshop on Intelligent Transportation)*, 2007.

[GG07]   M. Gerlach and F. Guttler. Privacy in VANETs using Changing Pseudonyms—Ideal and Real. In *65th IEEE Semiannual Vehicular Technology Conference, VTC 2007 Spring, Dublin, Ireland, April 22 – 25, 2007*, pages 2521–2525. IEEE Press, 2007.

[GGS04]   P. Golle, D. Greene, and J. Staddon. Detecting and Correcting Malicious Data in VANETs. In *VANET '04: Proceedings of the 1st International Workshop on Vehicular Ad-hoc Networks*, pages 29–37. ACM Press, 2004.

[GHOP01]   Michael Geber, Jürgen Hubrig, Jörn-Marten Ohle, and Andreas Pohlmann. Ausbausicherung für elektronische Komponenten bei einem Kraftfahrzeug. German Patent DE-10021811-A1, November 2001.

[Gie05]   Gieschen Consultancy. Report: IP theft up 22%, massive $3 trillion counterfeits. *www.bascap.com*, May 2005.

[GLLD05]   Dimitris Gritzalis, Costas Lambrinoudakis, Dimitrios Lekkas, and S. Deftereos. Technical Guidelines for Enhancing Privacy and Data Protection in Modern Electronic Medical Environments. *IEEE Transactions on Information Technology in Biomedicine*, 9(3):413–423, 2005.

[GPW⁺04]   Nils Gura, Arun Patel, Arvinderpal Wander, Hans Eberle, and Sheueling Chang Shantz. Comparing Elliptic Curve Cryptography and RSA on 8-bit CPUs. In *Cryptographic Hardware and Embedded Systems – CHES 2004*, volume 3156 of *LNCS*, pages 119–132. Springer-Verlag, 2004.

[GST05]   GST-SEC: Security in Global Systems for Telematics. *www.gstproject.org/sec/*, 2005.

[GVP+03]   Prasanth Ganesan, Ramnath Venugopalan, Pushkin Peddabachagari, Alexander Dean, Frank Mueller, and Mihail Sichitiu. Analyzing and Modeling Encryption Overhead for Sensor Network Nodes. In *WSNA '03: Proceedings of the 2nd ACM International Conference on Wireless Sensor Networks and Applications*, pages 151–159. ACM Press, 2003.

[GvW03]   Mark G. Graff and Kenneth R. van Wyk. *Secure Coding: Principles & Practices*. O'Reilly & Associates, Inc., 2003.

[Hås85]   Johan Håstad. On Using RSA with Low Exponent in a Public Key Network. In *Advances in Cryptology—CRYPTO '85*, volume 218 of *LNCS*, pages 403–408. Springer-Verlag, 1985.

[HC06]   Taimur Hassan and Samir Chatterjee. A Taxonomy for RFID. In *Proceedings of the 39th Annual Hawaii International Conference on System Sciences (HICSS'06)*. IEEE Computer Society, 2006.

[HCL04]   Jean-Pierre Hubaux, Srdjan Capkun, and Jun Luo. The Security and Privacy of Smart Vehicles. *IEEE Security & Privacy Magazine*, 2(3):49–55, 2004.

[HFPS02]   R. Housley, W. Ford, W. Polk, and D. Solo. RFC 3280: Internet X.509 Public Key Infrastructure Certificate and Certificate Revocation List (CRL) Profile. Technical report, RSA Laboratories, NIST, VeriSign, and Citigroup, *www.ietf.org/rfc/rfc3280.txt*, April 2002.

[Hil92]   Dan Hildebrand. An Architectural Overview of QNX. In *Proceedings of the Workshop on Micro-kernels and Other Kernel Architectures*, pages 113–126. USENIX Association, 1992.

[HIS04]   HIS – Herstellerinitiative Software. *www.automotive-his.de*, 2004.

[HJP03]   Yih-Chun Hu, David B. Johnson, and Adrian Perrig. SEAD: Secure Efficient Distance Vector Routing for Mobile Wireless Ad hoc Networks. *Ad Hoc Networks*, 1(1):175–192, 2003.

[HL03]   Michael Howard and David LeBlanc. *Writing Secure Code*. Microsoft Press, 2003.

[HM04]   Greg Hoglund and Gary McGraw. *Exploiting Software: How to Break Code*. Addison-Wesley, 2004.

[HMV04]   Darrel Hankerson, Alfred J. Menezes, and Scott A. Vanstone. *Guide to Elliptic Curve Cryptography*. Springer-Verlag, 2004.

[HPH98]   Jeffrey Hoffstein, Jill Pipher, and Joseph H.Silverman. NTRU: A Ring-Based Public Key Cryptosystem. In *Algorithmic Number Theory, Third International Symposium, ANTS-III, Portland, Oregon, USA, June 21 – 25, 1998*, volume 1423 of *LNCS*, pages 267–288. Springer-Verlag, 1998.

[HPJ02]   Yih-chun Hu, Adrian Perrig, and David B. Johnson. Wormhole Detection in Wireless Ad Hoc Networks. Technical Report TR01-384, Department of Computer Science, Rice University, June 2002.

[HPWW05] Katrin Höper, Christof Paar, André Weimerskirch, and Marko Wolf. Cryptographic Component Identification: Enabler for Secure Vehicles. In *62nd IEEE Semiannual Vehicular Technology Conference, VTC 2005 Fall, Dallas, Texas, USA, September 25 – 28, 2005*. IEEE Press, 2005.

[HRS98]   Kirsten M. Hansen, Anders P. Ravn, and Victoria Stavridou. From Safety Analysis to Software Requirements. *IEEE Transactions on Software Engineering*, 24(7):573–584, 1998.

[HSB⁺02]   H. Heinecke, A. Schedl, J. Berwanger, M. Peller, V. Nieten, R. Belschner, B. Hedenetz, P. Lohrmann, and C. Bracklo. FlexRay – ein Kommunikationssystem für das Automobil der Zukunft. Elektronik Automotive 09/2002, WEKA Fachzeitschriften-Verlag, 2002.

[HSW06]   Ulrich Huber, Ahmad-Reza Sadeghi, and Marko Wolf. Security Architectures for Software Updates and Content Protection. In *Automotive – Safety & Security 2006, Stuttgart, Germany, October 11 – 13, 2006*. Gesellschaft für Informatik e.V. (GI), Shaker-Verlag, 2006.

[HT98]   Günter Heiner and Thomas Thurner. Time-Triggered Architecture for Safety-Related Distributed Real-Time Systems in Transportation Systems. In *The 28th International Symposium on Fault-Tolerant Computing (FTCS 1998)*, pages 402–407, 1998.

[IBM08]   IBM Corporation. Rational PurifyPlus. *www.ibm.com/software/ awdtools/purifyplus/*, 2008.

[IEE00]   IEEE P1363-2000. *Standard Specifications for Public-Key Cryptography*. IEEE, 2000.

[IEE04]   IEEE P1363a-2004. *Standard Specifications for Public-Key Cryptography – Amendment 1: Additional Techniques*. IEEE, 2004.

[IEE05]   IEEE 1616-2004. *Standard for Motor Vehicle Event Data Recorders (MVEDRs)*. IEEE, 2005.

[IEE06a]   IEEE 1609-2006. *Family of Standards for Wireless Access in Vehicular Environments (WAVE)*. IEEE, 2006.

[IEE06b]   IEEE P1609.2-2006. *Standard for Wireless Access in Vehicular Environments (WAVE) – Security Services for Applications and Management Messages*. IEEE, 2006.

[IEE07]   IEEE 802.11x. *Standards for Wireless Local Area Networks (WLANs)*. IEEE, 2007.

[Int04]   The Intelligent Transportation (WIT) Workshop Series. *http://wit.tu-harburg.de*, 2004.

[Int07]   Intel Corporation. Intel Trusted Execution Technology – Preliminary Architecture Specification and Enabling Considerations. Technical Report 31516804, Intel Corporation, August 2007.

[ISO03]   ISO/IEC 10007:2003. *Quality Management Systems – Guidelines for Configuration Management*. ISO/IEC, 2003.

[ISO04a]   ISO/IEC 13335:2004. *Management of Information and Communications Technology Security*. ISO/IEC, 2004.

[ISO04b]   ISO/IEC 15446:2004. *Information Technology – Security Techniques – Guide for the Production of Protection Profiles and Security Targets*. ISO/IEC, 2004.

[ISO05a]   ISO/IEC 15408:2005. *Information Technology – Security Techniques – Evaluation Criteria for IT Security*. ISO/IEC, 2005.

[ISO05b]   ISO/IEC 15443:2005. *Information Technology – Security Techniques – A Framework for IT security Assurance*. ISO/IEC, 2005.

[ISO05c]   ISO/IEC 27001:2005. *Information technology – Security Techniques – Specification for an Information Security Management System*. ISO/IEC, 2005.

[ISO06]   ISO/IEC 11898:2006. *Road Vehicles – Controller Area Network (CAN)*. ISO/IEC, 2006.

[ISO07a]   ISO/IEC 13491:2007. *Banking – Secure Cryptographic Devices.* ISO/IEC, 2007.

[ISO07b]   ISO/IEC 27002:2005. *Information Technology – Security Techniques – Code of Practice for Information Security Management.* ISO/IEC, 2007.

[Jap08]   The Japan Automotive Software Platform Architecture (JasPar) Consortium. *www.jaspar.jp*, 2008.

[JBP+97]   Bjorn Markus Jakobsson, Mihir Bellare, Ramamohan Paturi, Nolan Wallach, Bennet Yee, and Moti Yung. *Privacy vs. Authenticity*. PhD thesis, University of California, 1997.

[JBR99]   Ivar Jacobson, Grady Booch, and Jim Rumbaugh. *The Unified Software Development Process*. Addison-Wesley, 1999.

[JG97]   Roger G. Johnston and Anthony R.E. Garcia. Vulnerability Assessment of Security Seals. Technical Report LA-UR-96-3672, Los Alamos National Laboratory, January 1997.

[JG07]   M. Eric Johnson and Eric Goetz. Embedding Information Security into the Organization. *IEEE Security and Privacy*, 5(3):16–24, 2007.

[JK02]   J. Jonsson and B. Kaliski. RFC 3447: Public-Key Cryptography Standards (PKCS) #1: RSA Cryptography. Technical report, RSA Laboratories Inc., *www.ietf.org/rfc/rfc3447.txt*, June 2002.

[Joh04]   Eric Johnson. The Safety of Secrets in Extended Enterprises. The Financial Times, August 18, 2004.

[Kau06]   Minderjeet Kaur. 'MyKad' for Cars to Help Stop Thieves. The New Straits Times, December 9, 2006.

[KK99]   Oliver Kömmerling and Markus G. Kuhn. Design Principles for Tamper-Resistant Smartcard Processors. In *Proceedings of the USENIX Workshop on Smartcard Technology (SMARTCARD-99)*, pages 9–20. USENIX Association, 1999.

[KLM⁺04]   Paul Kocher, Ruby B. Lee, Gary McGraw, Anand Raghunathan, and
            Srivaths Ravi. Security as a New Dimension in Embedded System
            Design. In *Fortyfirst Design Automation Conference, DAC 2004*,
            pages 753–760. IEEE Press, 2004.

[KMS06]   Frank Kargl, Zhendong Ma, and Elmar Schoch. Security Engineer-
            ing for VANETs. In *4th Workshop on Embedded Security in Cars,
            escar 2006, Bochum, Germany, November 14 – 15*, 2006.

[Kob87]   Neal Koblitz. Elliptic Curve Cryptosytems. *Mathematics of Com-
            putation*, 48(177):203–209, 1987.

[Koo08]   Bert-Jaap Koops. Survey of Cryptography Laws. *http://rechten.uvt.
            nl/koops/cryptolaw/*, 2008.

[Kow01]   T.M. Kowalick. Pros and Cons of Emerging Event Data Recorders
            (EDRs) in the Highway Mode of Transportation. In *53rd IEEE
            Semiannual Vehicular Technology Conference, VTC 2001 Spring,
            Rhodes, Greece, May 6 – 9, 2001*. IEEE Press, 2001.

[KPP⁺06]   Sandeep Kumar, Christof Paar, Jan Pelzl, Gerd Pfeiffer, Andy Rupp,
            and Manfred Schimmler. How to Break DES for Euro 8,980.
            In *Special-purpose Hardware for Attacking Cryptographic Systems
            (SHARCS 2006), Cologne, Germany, April 3 – 4*, 2006.

[Kuh04]   Markus G. Kuhn. An Asymmetric Security Mechanism for Naviga-
            tion Signals. In *6th International Workshop on Information Hiding,
            IH 2004, Toronto, Canada, May 23 – 25, 2004*. Springer-Verlag,
            2004.

[Kuh06]   Markus G. Kuhn. Positioning Security – From Electronic Warfare
            to Cheating RFID and Road-Tax Systems. In *4th Workshop on Em-
            bedded Security in Cars, escar 2006, Bochum, Germany, November
            14 – 15*, 2006.

[Lar05]   Ola Larses. Factors Influencing Dependable Modular Architec-
            tures for Automotive Applications. Technical Report TRITA-MMK
            2005:09, Mechatronics Lab, Department of Machine Design, Royal
            Institute of Technology, KTH Stockholm, March 2005.

[LB07]   Christine Laurendeau and Michel Barbeau.   Secure Anonymous
            Broadcasting in Vehicular Networks. In *IEEE Conference on Lo-
            cal Computer Networks (LCN 2007)*, pages 661–668. IEEE Press,
            2007.

[LD04]   Andreas Lang and Jana Dittmann. Steigende Informationstechnolo-
            gie: Sicherheitsrisiko im Fahrzeugbau. In *Automotive – Safety & Se-
            curity 2004, Stuttgart, Germany, October 6 – 7, 2004*. Gesellschaft
            für Informatik e.V. (GI), Shaker-Verlag, 2004.

[Lem06]   Kerstin Lemke. Physical Protection against Tampering Attacks. In
            *Embedded Security in Cars: Securing Current and Future Automo-
            tive IT Applications*, pages 207–217. Springer-Verlag, 2006.

[LIN99]   The LIN Consortium. *www.lin-subbus.org*, 1999.

[LL07]   Jie Liang and Xue-Jia Lai. Improved Collision Attack on Hash Func-
            tion MD5. *Journal of Computer Science and Technology*, 22(1):79–
            87, 2007.

[LPW06]   Kerstin Lemke, Christof Paar, and Marko Wolf, editors. *Embedded
            Security in Cars: Securing Current and Future Automotive IT Ap-
            plications*. Springer-Verlag, 2006.

[LR07]   Kerstin Lemke-Rust. *Models and Algorithms for Physical Crypt-
            analysis*. Europäischer Universitätsverlag, 2007.

[LSK06]   Tim Leinmüller, Elmar Schoch, and Frank Kargl. Position Verifi-
            cation Approaches for Vehicular Ad Hoc Networks. *IEEE Wireless
            Communications, Special Issue on Inter-Vehicular Communications*,
            13(5):16–21, 2006.

[LSS05]   Kerstin Lemke, Ahmad-Reza Sadeghi, and Christian Stüble. An
            Open Approach for Designing Secure Electronic Immobilizers. In
            *Information Security Practice and Experience Conference, ISPEC
            2005, Singapore, April 11-14, 2005*, volume 3439 of *LNCS*, pages
            230–242. Springer-Verlag, 2005.

[McG06]   Gary McGraw. *Software Security: Building Security In*. Addison-
            Wesley, 2006.

[Mic08a]   Microchip Technology Inc. The KEELOQ Authentication. *www.
            microchip.com*, 2008.

[Mic08b]   Microsoft Corporation. BitLocker Drive Encryption. *http://www. microsoft.com/technet/windowsvista/security/bitlockr.mspx*, 2008.

[Mil85]   Victor S. Miller. Use of Elliptic Curves in Cryptography. In *Advances in Cryptology – CRYPTO '85*, volume 218 of *LNCS*, pages 417–426. Springer-Verlag, 1985.

[MKK$^+$06]   Thomas Miehling, Burkhard Kuhls, Heiko Kober, Hartmut Chodura, and Marcus Heitmann. HIS Security Module. Specification Version 1.1, Herstellerinitiative Software (HIS), 2006.

[MLO97]   Terrence Mitchem, Raymond Lu, and Richard O'Brien. Using Kernel Hypervisors to Secure Applications. In *Proceedings of 13th Annual Computer Security Applications Conference (ACSAC)*, pages 175–181. IEEE Press, 1997.

[MMJ05]   Milena Milenković, Aleksandar Milenković, and Emil Jovanov. Hardware Support for Code Integrity in Embedded Processors. In *Proceedings of the 2005 International Conference on Compilers, Architectures and Synthesis for Embedded Systems, CASES 2005*, pages 55–65. ACM Press, 2005.

[MN03]   Jonathan D. Moffett and Bashar A. Nuseibeh. A Framework for Security Requirements Engineering. Technical Report YCS 368, University of York, UK, August 2003.

[Moo65]   Gordon E. Moore. Cramming more Components onto Integrated Circuits. *Electronics*, 38(8):114–117, 1965.

[Moo90]   A. Moon. Vehicle Control Systems – Reliability through Simplicity. In *IEE Colloquium on Safety-Critical Software in Vehicle & Traffic Control*. Institution of Electrical Engineers (IEE), London, UK, 1990.

[MOS98]   The MOST Cooperation. *www.mostcooperation.com*, 1998.

[MS07]   Christoph Marscholik and Peter Subke. *Datenkommunikation im Automobil. Grundlagen, Bussysteme, Protokolle und Anwendungen.* Hüthig-Verlag, 2007.

[MVH$^+$06]   Thomas Mieling, Pavel Vondracek, Martin Huber, Hartmut Chodura, and Gerhard Bauersachs. HIS Flashloader. Specification Version 1.1, Herstellerinitiative Software (HIS), 2006.

[MvOV96]   Alfred J. Menezes, Paul C. van Oorschot, and Scott A. Vanston.
           Handbook of Applied Cryptography. CRC Press, 1996.

[MZ05]   Norman Meyersohn and Tom Zeller. Can a Virus Hitch a Ride in
           Your Car? The New York Times, March 13, 2005.

[Nat08]   National Institute of Standards and Technology. The Cryptographic
           Hash Project. *http://csrc.nist.gov/groups/ST/hash/*, 2008.

[Net04]   The Network on Wheels (NOW) Project. *www.network-on-wheels.
           de*, 2004.

[NHR05]   C. Neuman, S. Hartman, and K. Raeburn. RFC 4120: The Kerberos
           Network Authentication Service (V5). Technical report, Project
           Athena, MIT, *www.ietf.org/rfc/rfc4120.txt*, July 2005.

[NIS07]   NIST Special Publication 800-57. *Recommendation for Key Man-
           agement.* National Institute of Standards and Technology, 2007.

[Nor08]   Norwich Union. The "Pay As You Drive" Insurance. *www.
           norwichunion.com/pay-as-you-drive/*, 2008.

[Off06]   Office of Regulatory Analysis and Evaluation. Event Data Recorders
           (EDRs). Final regulatory evaluation, National Center for Statistics
           and Analysis, July 2006.

[OnS08]   OnStar Corporation. *www.onstar.com*, 2008.

[Ope08]   The OpenSSL Project. *www.openssl.org*, 2008.

[OSE08]   OSEK/VDX Group. The OSEK Operating System. *www.osek-vdx.
           org*, 2008.

[Paa03]   Christof Paar. Eingebettete Sicherheit im Automobil. In *1st Work-
           shop on Embedded Security in Cars, escar 2003, Bochum, Germany,
           November 18 – 19*, 2003.

[Pat99]   Andy Patrizio. Why the DVD Hack Was a Cinch. *Wired News*, 2,
           1999.

[PB61]   William Wesley Peterson and D.T. Brown. Cyclic Codes for Error
           Detection. *Proceedings of the IRE*, 49(1):228–235, 1961.

[PBH+07]   P. Papadimitratos, L. Buttyan, J.-P. Hubaux, F. Kargl, A. Kung, and M. Raya. Architecture for Secure and Private Vehicular Communications. In *7th International Conference on ITS Telecommunications (ITST 2007)*, pages 1–6, 2007.

[PGH06]   Panagiotis Papadimitratos, Virgil Gligor, and Jean-Pierre Hubaux. Securing Vehicular Communications – Assumptions, Requirements, and Principles. In *4th Workshop on Embedded Security in Cars, escar 2006, Bochum, Germany, November 14 – 15*, 2006.

[Pol95]   Stefan Poledna. Fault Tolerance in Safety Critical Automotive Applications: Cost of Agreement as a Limiting Factor. In *The 25th International Symposium on Fault-Tolerant Computing (FTCS 1995)*, pages 73–82, 1995.

[PP05]   Bryan Parno and Adrian Perrig. Challenges in Securing Vehicular Networks. In *Workshop on Hot Topics in Networks (HotNets-IV)*, 2005.

[PRe04]   The PReVENTive and Active Safety Applications Integrated Project (PReVENT IP). *www.prevent-ip.org*, 2004.

[Pri08]   The Privacy Rights Clearinghouse. A Chronology of Data Breaches. *www.privacyrights.org/ar/ChronDataBreaches.htm*, 2008.

[PRS+01]   Birgit Pfitzmann, James Riordan, Christian Stüble, Michael Waidner, and Arnd Weber. The PERSEUS System Architecture. Technical Report RZ 3335 (#93381), IBM Research Division, Zurich Laboratory, April 2001.

[PSS+04]   M. Peden, R. Scurfield, D. Sleet, D. Mohan, A. Hyder, E. Jarawan, and C. Mathers. *World Report on Road Traffic Injury Prevention*. World Health Organization (WHO), 2004.

[PT06]   Michael Pecht and Sanjay Tiku. Bogus! IEEE Spectrum, May, 2006.

[Puc01]   Heike Puchan. The Mercedes-Benz A-class Crisis. *Corporate Communications: An International Journal*, 6(1):42–46, 2001.

[PW06]   Jan Pelzl and Thomas Wollinger. Security Aspects of Mobile Communication Systems. In *Embedded Security in Cars: Securing Current and Future Automotive IT Applications*, pages 167–185. Springer-Verlag, 2006.

[PW08]   Christof Paar and Marko Wolf. Security Requirements Engineering
            in the Automotive Domain: On Specification Procedures and Im-
            plementational Aspects. In *SICHERHEIT 2008: Sicherheit - Schutz
            und Zuverlässigkeit, Beiträge der 4. Jahrestagung des Fachbereichs
            Sicherheit der Gesellschaft für Informatik e.V. (GI), Saarbrücken,
            Germany, April 2 – 4, 2008*, volume 110 of *LNI*. Gesellschaft für
            Informatik e.V. (GI), Köllen-Verlag, 2008.

[PWW04a]   Christof Paar, André Weimerskirch, and Marko Wolf. Komponen-
            tenidentifikation: Voraussetzung für IT-Sicherheit im Automobil. In
            *Automotive – Safety & Security 2004, Stuttgart, Germany, October
            6 – 7, 2004*. Gesellschaft für Informatik e.V. (GI), Shaker-Verlag,
            2004.

[PWW04b]   Christof Paar, André Weimerskirch, and Marko Wolf. Security in
            Automotive Bus Systems. In *2nd Workshop on Embedded Security
            in Cars, escar 2004, Bochum, Germany, November 10 – 11*, 2004.

[PWW04c]   Christof Paar, André Weimerskirch, and Marko Wolf. Sicherheit in
            automobilen Bussystemen. In *Automotive – Safety & Security 2004,
            Stuttgart, Germany, October 6 – 7, 2004*. Gesellschaft für Informatik
            e.V. (GI), Shaker-Verlag, 2004.

[PWW05]   Christof Paar, André Weimerskirch, and Marko Wolf. Digital Rights
            Management Systeme (DRMS) als *Enabling Technology* im Au-
            tomobil. In *SICHERHEIT 2005: Sicherheit - Schutz und Zuver-
            lässigkeit, Beiträge der 2. Jahrestagung des Fachbereichs Sicher-
            heit der Gesellschaft für Informatik e.V. (GI), Regensburg, Germany,
            April 5 – 8, 2005*, volume 62 of *LNI*. Gesellschaft für Informatik e.V.
            (GI), Köllen-Verlag, 2005.

[RAH06]   Maxim Raya, Adel Aziz, and Jean-Pierre Hubaux. Efficient Secure
            Aggregation in VANETs. In *VANET '06: Proceedings of the 3rd
            International Workshop on Vehicular Ad-hoc Networks*, pages 67–
            75. ACM Press, 2006.

[RCCL06]   C. L. Robinson, L. Caminiti, D. Caveney, and K. Laberteaux. Ef-
            ficient Coordination and Transmission of Data for Cooperative Ve-
            hicular Safety Applications. In *VANET '06: Proceedings of the 3rd
            International Workshop on Vehicular Ad-hoc Networks*, pages 10–
            19. ACM Press, 2006.

[RG06]   Klaus Rüdiger and Martin Gersch. In-Vehicle M-Commerce: Busi-
             ness Models for Navigation Systems and Location-based Services.
             In *Embedded Security in Cars: Securing Current and Future Auto-
             motive IT Applications*, pages 247–273. Springer-Verlag, 2006.

[RH07]   Maxim Raya and Jean-Pierre Hubaux. Securing Vehicular Ad Hoc
             Networks. *Journal of Computer Security, Special Issue on Security
             of Ad Hoc and Sensor Networks*, 15(1):39–68, 2007.

[Riv92]   Ron Rivest. RFC 1321: The MD5 Message-Digest Algorithm. Tech-
             nical report, MIT Laboratory for Computer Science and RSA Data
             Security, Inc., *www.ietf.org/rfc/rfc1321.txt*, April 1992.

[RKH04]   Anand Ravi, Srivathsand Raghunathan, Paul Kocher, and Sunil
             Hattangady. Security in Embedded Systems: Design Challenges.
             ACM Transactions on Embedded Computing Systems, 3(3):461–
             491, 2004.

[Ros04]   Sativa Ross. Parts Counterfeiting. *www.aftermarketbusiness.com*,
             October 2004.

[RPH06]   Maxim Raya, Panos Papadimitratos, and Jean-Pierre Hubaux. Se-
             curing Vehicular Communications. *IEEE Wireless Communica-
             tions Magazine, Special Issue on Inter-Vehicular Communications*,
             13(5):8–15, 2006.

[RSA78]   R. L. Rivest, A. Shamir, and L. Adleman. A Method for Obtaining
             Digital Signatures and Public-Key Cryptosystems. *Communications
             of the ACM*, 21(2):120–126, 1978.

[Rud03]   Michael Rudorfer. IT-Trends im Fahrzeug. In *1st Workshop on Em-
             bedded Security in Cars, escar 2003, Bochum, Germany, November
             18 – 19*, 2003.

[Rus01]   John Rushby. Security Requirements Specifications: How and
             What? In *Symposium on Requirements Engineering for Informa-
             tion Security (SREIS)*, 2001.

[S+02]   Jochem Spohr et al. Requirements for Protection of Applications
             under OSEK. Technical report, Herstellerinitiative Software (HIS),
             2002.

[S⁺08]   Julian Seward et al. Valgrind – Debugging and Profiling Linux Pro-
            grams. *www.valgrind.org*, 2008.

[SAF06]   The SAFESPOT Project. Cooperative Vehicles and Road Infrastruc-
            ture for Road Safety. *www.safespot-eu.org*, 2006.

[SB07]   William Stallings and Lawrie Brown. *Computer Security: Princi-
            ples and Practice*. Prentice-Hall, Inc., 2007.

[Sch96]   Bruce Schneier. *Applied Cryptography*, chapter 15 – Combining
            Ciphers. John Wiley & Sons, 1996.

[Sch06]   Kai Schramm. *Advanced Methods in Side Channel Cryptanalysis*.
            Europäischer Universitätsverlag, 2006.

[SeV06]   The SeVeCom Project. Secure Vehicular Communication. *www.
            sevecom.org*, 2006.

[Sha49]   Claude Shannon. Communication Theory of Secrecy Systems. *The
            Bell System Technical Journal*, 28(74):656–715, 1949.

[Sha79]   Adi Shamir. How to Share a Secret. *Communications of the ACM*,
            22(11):612–613, 1979.

[Shi00]   R. Shirley. RFC 2828: Internet Security Glossary. Technical report,
            GTE/BBN Technologies, *www.ietf.org/rfc/rfc2828.txt*, May 2000.

[SHL⁺05]   Krishna Sampigethaya, Leping Huang, Mingyan Li, Radha Pooven-
            dran, Kanta Matsuura, and Kaoru Sezaki. CARAVAN: Providing
            Location Privacy for VANET. In *3rd Workshop on Embedded Se-
            curity in Cars, escar 2005, Bochum, Germany, November 29 – 30*,
            2005.

[Sho97]   Peter W. Shor. Polynomial-Time Algorithms for Prime Factorization
            and Discrete Logarithms on a Quantum Computer. *SIAM Journal on
            Computing*, 26(5):1484–1509, 1997.

[Sie07]   Siemens VDO Automotive AG. New Dimensions in Driving Com-
            fort and Safety. pro.pilot – Advanced Driver Assistance Systems.
            White paper, Siemens VDO Automotive AG, 2007.

[Sin98]   Abraham Sinkov. *Elementary Cryptanalysis: A Mathematical Ap-
            proach*. The Mathematical Association of America, 1998.

[SJV$^+$05]   Reiner Sailer, Trent Jaeger, Enriquillo Valdez, Ramon Caceres, Ronald Perez, Stefan Berger, John Linwood Griffin, and Leendert van Doorn. Building a MAC-Based Security Architecture for the Xen Open-Source Hypervisor. In *Proceedings of the 21st Annual Computer Security Applications Conference (ACSAC)*, pages 276–285. IEEE Press, 2005.

[SKC00]   Stephan Schmitz, Jacek Kruppa, and Peter Crowhurst. Safety and Security Considerations of New Closure Systems. Technical Paper 2000-01-1304, SAE International, March 2000.

[SLLG05]   Weidong Shi, Hsien-Hsin S. Lee, Chenghuai Lu, and Mrinmoy Ghosh. Towards the Issues in Architectural Support for Protection of Software Execution. *ACM SIGARCH Computer Architecture News*, 33(1):6–15, 2005.

[SLP06]   Kai Schramm, Kerstin Lemke, and Christof Paar. Side Channel Attacks. In *Embedded Security in Cars: Securing Current and Future Automotive IT Applications*, pages 187–206. Springer-Verlag, 2006.

[Sma00]   Nigel P. Smart. Physical Side-channel Attacks on Cryptographic Systems. *Software Focus*, 1(2):6–13, 2000.

[Son06]   Wes Sonnenreich. Return On Security Investment (ROSI): A Practical Quantitative Model. *Journal of Research and Practice in Information Technology*, 38(1):55–66, 2006.

[Spe06]   Ed Sperling. IP Theft Continues. *www.edn.com/article/CA6305688. html*, Electronic News, February 2, 2006.

[SPvDK04]   Arvind Seshadri, Adrian Perrig, Leendert van Doorn, and Pradeep Khosla. Using Software-based Attestation for Verifying Embedded Software in Cars. In *2nd Workshop on Embedded Security in Cars, escar 2004, Bochum, Germany, November 10 – 11*, 2004.

[SRM06]   Winfried Stephan, Solveig Richter, and Markus Müller. Aspects of Secure Vehicle Software Flashing. In *Embedded Security in Cars: Securing Current and Future Automotive IT Applications*, pages 17–26. Springer-Verlag, 2006.

[SS08]   Marcel Selhorst and Christian Stüble. Trusted GRUB 1.1.3. *http: //sourceforge.net/projects/trustedgrub/*, 2008.

[SSS+06]   Ahmad-Reza Sadeghi, Michael Scheibel, Oskar Senft, Christian Stüble, Marcel Winandy, and Marko Wolf. Design and Implementation of a Secure Linux Device Encryption Architecture. Research report, Horst Görtz Institute for IT Security, June 2006.

[SSSW06]   Ahmad-Reza Sadeghi, Michael Scheibel, Christian Stüble, and Marko Wolf. Play it once again, Sam – Enforcing Stateful Licenses on Open Platforms. In *2nd Workshop on Advances in Trusted Computing, WATC 2006 Fall, IBM Tokyo Research Labs, Tokyo, Japan, November 30 – December 1, 2006*, pages 63–78, 2006.

[SSW06]   Michael Scheibel, Christian Stüble, and Marko Wolf. Design and Implementation of an Architecture for Vehicular Software Protection. In *4th Workshop on Embedded Security in Cars, escar 2006, Bochum, Germany, November 14 – 15*, 2006.

[SSW08]   Michael Scheibel, Christian Stüble, and Marko Wolf. An Interoperable Security Architecture for Vehicular Software Protection. In *International Workshop on Interoperable Vehicles, IOV 2008, ETH Zürich, Switzerland, March 26, 2008*, pages 72–78, 2008.

[Sta98]   The Standards for Efficient Cryptography Group (SECG). *www.secg.org*, 1998.

[Sti95]   Douglas R. Stinson. *Cryptography: Theory and Practice*. CRC Press, 1995.

[STM07]   STMicroelectronics Inc. Q2 2007 Financial Results. *investors.st.com*, 2007.

[SV05]   Alberto Sangiovanni-Vincentelli. Integrated Electronics in the Car and the Design Chain Evolution or Revolution? In *DATE '05: Proceedings of the conference on Design, Automation and Test in Europe*, pages 532–533. IEEE Press, 2005.

[SW03]   Alexandre Saad and Ulrich Weinmann. Automotive Software Engineering and Concepts. In *INFORMATIK 2003 – Innovative Informatikanwendungen, Band 1, Beiträge der 33. Jahrestagung der Gesellschaft für Informatik e.V. (GI), 29. September - 2. Oktober 2003 in Frankfurt am Main*, volume 34 of *LNI*, pages 318–319. GI, 2003.

[SYS08]   SYSGO AG. The PikeOS RTOS Technology. *www.sysgo.com*,
            2008.

[TB06]   Pim Tuyls and Lejla Batina. RFID-Tags for Anti-counterfeiting. In
            *Proceedings of the Cryptographers' Track at the RSA Conference
            (CT-RSA 2006)*, volume 3860 of *LNCS*, pages 115–131. Springer-
            Verlag, 2006.

[Tol08]   Toll Collect GmbH. *www.toll-collect.de*, 2008.

[Tru03]   Trusted Computing Group (TCG). *www.trustedcomputinggroup.
            org*, 2003.

[Tru05]   Trusted Computing Group (TCG). Embedded Systems and Trusted
            Computing Security. White paper, TPM Work Group, September
            2005.

[Tru06]   Trusted Computing Group (TCG). Securing Mobile Devices on
            Converged Networks. White paper, Mobile Phone Work Group,
            September 2006.

[Tru07a]   Trusted Computing Group (TCG). TCG Software Stack (TSS).
            Specification Version 1.2, Level 1, Errata A, TPM Software Stack
            Work Group, March 2007.

[Tru07b]   Trusted Computing Group (TCG). Trusted Platform Module (TPM).
            Main Specification Version 1.2 Revision 103, TPM Work Group,
            July 2007.

[TWP07]   Erik Tews, Ralf-Philipp Weinmann, and Andrei Pyshkin. Breaking
            104 bit WEP in less than 60 seconds. Cryptology eprint Archive,
            Report 2007/120, 2007.

[Ulk00]   Walter Ulke. Remote Entry Control and Immobilizer System So-
            lutions. Technical Paper 2000-01-1310, SAE International, March
            2000.

[US-06]   US-CFR-49-563. *United States Code of Federal Regulations Title
            49 Part 563 on Event Data Recorders*. United States Department of
            Transportation – National Highway Traffic Safety Administration,
            2006.

[vBC05]   Gerard van Battum and Dario Carluccio. Physical Security for Au-
            tomotive Applications: What can we learn from other industries? In
            *3rd Workshop on Embedded Security in Cars, escar 2005, Bochum,
            Germany, November 29 – 30*, 2005.

[Veh50]   The IEEE Vehicular Technology (VTC) Conference Series. *www.
            ieeevtc.org*, 1950.

[Veh04]   The Vehicular Ad Hoc Networks (VANET) Workshop Series. *www.
            path.berkeley.edu/ vanet/*, 2004.

[Veh05]   The Vehicle to Vehicle Communications (V2VCOM) Workshop Se-
            ries. *www.v2vcom.org*, 2005.

[Ver26]   G. S. Vernam. Cipher Printing Telegraph Systems for Secret Wire
            and Radio Telegraphic Communications. *Journal of the American
            Institute of Electrical Engineers*, 55:109–115, 1926.

[VM02]   John Viega and Gary McGraw. *Building Secure Software: How to
            Avoid Security Problems the Right Way*. Addison-Wesley, 2002.

[vO03]   Paul C. van Oorschot. Revisiting Software Protection. In *6th In-
            ternational Information Security Conference (ISC)*, volume 2851 of
            LNCS, pages 1–13. Springer-Verlag, 2003.

[Was04]   Lawrence C. Washington. *Elliptic Curves: Number Theory and
            Cryptography*. Chapman & Hall/CRC Press, 2004.

[Wei00]   Steve Weingart. Physical Security Devices for Computer Subsys-
            tems: Survey of Attacks and Defenses. In *Cryptographic Hardware
            and Embedded Systems – CHES 2000*, volume 1965 of *LNCS*, pages
            302–317. Springer-Verlag, 2000.

[WFM⁺07]   Peter Wilson, Alexandre Frey, Tom Mihm, Danny Kershaw, and
            Tiago Alves. Implementing Embedded Security on Dual-Virtual-
            CPU Systems. *IEEE Design & Test of Computers*, 24(6):582–591,
            2007.

[Whe08]   David A. Wheeler. Flawfinder. *www.dwheeler.com/ flawfinder/*,
            2008.

[Wik08]   Wikipedia, The Free Encyclopedia. Block Cipher Modes of
            Operation. *http://en.wikipedia.org/ wiki/ Block_cipher_modes_of_
            operation*, 2008.

[Win08]   Wind River Systems Inc. The VxWorks RTOS. *wwww.windriver.*
            com/vxworks/, 2008.

[WJ03]   Jon S. Warner and Roger G. Johnston. Think GPS Cargo Tracking =
            High Security? Think Again. Technical report, Los Alamos National
            Laboratory, September 2003.

[WWW06]   Christof Wegener, André Weimerskirch, and Marko Wolf. Rechte
            für Kleine – Digital Rights Management in mobilen und eingebet-
            teten Geräten. iX – Magazin für professionelle Informationstechnik
            1/2006, heise Zeitschriftenverlag, 2006.

[WWW07]   Marko Wolf, André Weimerskirch, and Thomas Wollinger. State of
            the Art: Embedding Security in Vehicles. *EURASIP Journal on Em-*
            bedded Systems (EURASIP JES), Special Issue: Embedded Systems
            for Intelligent Vehicles, 2007:Article ID 74706, 2007.

[YB97]   Victor Yodaiken and Michael Barabanov. A Real-Time Linux, 1997.

[ZH99]   Lidong Zhou and Zygmunt J. Haas. Securing Ad Hoc Networks.
            IEEE Network, 13(6):24–30, 1999.

[Zim05]   M. Zimmer. Surveillance, Privacy and the Ethics of Vehicle Safety
            Communication Technologies. *Ethics and Information Technology*,
            7(4):201–210, 2005.

[ZNK$^+$06] Michael Zweck, Anton Negele, Carsten Krebs, Martin Kolbe,
            Frank Scharberth, Albrecht Möckel, and Bernhard Gerdiken. HIS-
            konforme Programmierung von Steuergeräten auf Basis von UDS.
            Specification Version 1.0, Herstellerinitiative Software (HIS), 2006.

[ZP93]   E. Zanoni and P. Pavan. Improving the Reliability and Safety of
            Automotive Electronics. *IEEE Micro*, 13(10):30–48, 1993.

[ZS07]   Werner Zimmermann and Ralf Schmidgall. *Bussysteme in der*
            Fahrzeugtechnik. Vieweg-Verlag, 2007.

Index